Methods and Applications of Artificial Intelligence

Artificial Intelligence in Mechanics Set

coordinated by
Abdelkhalak El Hami

Volume 2

Methods and Applications of Artificial Intelligence

Dynamic Response, Learning, Random Forest, Linear Regression, Interoperability, Additive Manufacturing and Mechatronics

Edited by
Abdelkhalak El Hami

iSTE

WILEY

First published 2025 in Great Britain and the United States by ISTE Ltd and John Wiley & Sons, Inc.

ISTE Ltd
27-37 St George's Road
London SW19 4EU
UK

www.iste.co.uk

John Wiley & Sons, Inc.
111 River Street
Hoboken, NJ 07030
USA

www.wiley.com

Library of Congress Control Number: 2024949558

British Library Cataloguing-in-Publication Data
A CIP record for this book is available from the British Library
ISBN 978-1-78630-999-0

Contents

Preface

Artificial intelligence (AI) is currently one of the most discussed technologies, by both scientists and public media. Several factors have contributed to its development throughout the years. The first one is access to a large amount of data. In the industrial domain, it is the advent of Industry 4.0 that has promoted automation and data sharing in several technologies. An additional factor is the continuous enhancement of computational power due to the development of increasingly high-performance processors and algorithm optimization. These two limitations being removed, most developments in AI focus on the quality of predictions. Integrating AI in the industrial field is a new and exciting frontier for innovation.

Similar to the transformation brought about by AI in many other sectors, its application to technologies in mechanics offers the possibility to significantly improve design, manufacturing and quality control processes; from computer aided design (CAD) to the optimization of printing parameters, passing through fault detection and real-time monitoring. This type of technology requires information systems, data with management systems and advanced algorithms that can be used by AI.

In mechanics, AI offers several possibilities in engineering industry, predictive maintenance, monitoring of facilities, robotics, additive manufacturing, materials, vibrations, etc.

This book focuses on *Methods and Applications of Artificial Intelligence* in mechanics. It is organized into eight chapters.

The first chapter presents the dynamic response of a thermoplastic composite (PP-GF70) using an intelligent method. It uses *Crashworthiness* to redimension the size of future vehicles, by integrating composite materials lighter than steel that have a good mechanical and dynamic behavior. The work conducted in this chapter involves the comparison of the results of size-impact between an all-steel body and a body integrating a thermoplastic composite (PP-GF70). This material represents an innovative solution in several industries, thanks to these mechanical and thermal advantages and also to its manufacturing process. In the automotive industry, the race to save energy and to reduce pollutant gas emissions attracts an increasingly high interest of builders and equipment manufacturers for the use of low-density materials.

Chapter 2 takes a close look at the opportunities and challenges associated with AI integration into AM (Additive Manufacturing). It explores the various potential applications of AI in AM, such as topology optimization, prediction of printed part performances and automated management of manufacturing processes. Furthermore, we discuss the ethical considerations and safety implications of using AI in this context. This chapter explores the convergence of two dynamic domains: AI and AM, in order to address a critical issue of manufacturing. While AM evolves, it is increasingly clear that AI may play an essential role in the optimization and improvement of this manufacturing process. The objective is to explore the possibilities offered by AI in the context of additive manufacturing, by highlighting the potential advantages, the challenges and the research paths to be explored. Through this examination, this chapter aims to provide stimulating perspectives for the researchers and industry professionals, while contributing to a deeper understanding of the synergies between AI and AM.

Chapter 3 focuses on the reliability of electronic devices depending on the welded joints that connect the electronic components to the printed circuit boards. It is particularly essential for BGAs (Ball Grid Arrays), which integrate multiple contacts in a limited zone. It is in this context that this work has been conducted to evaluate the use of the design optimization method based on reliability and AI for optimization applications. It pairs the finite element model developed by the Ansys APDL software and the design optimization model based on the reliability coded with MATLAB software.

Chapter 4 presents the cooperative intelligent transportation systems (C-ITS) based on communications and sharing of information between vehicles, road infrastructures and communication infrastructure to improve users' safety, facilitate their travel and guarantee the safety of road workers. This chapter presents the technological advances in the field of ITS, particularly the democratization of the communication between vehicle and infrastructure using ITS-G5. A new electronic toll collection service based on this type of communication is presented. Its objective is to reduce maintenance costs, increase agents' safety and improve traffic fluidity.

The objective of *Chapter 5* is to propose an AI methodology for machine learning. It is used to help machines learn from data and make autonomous decisions. The random forest is an innovative approach in the field of machine learning, offering a robust and multi-purpose method for classification and regression. This exploration serves to discover the theoretical basis of this technique, its practical applications and the challenges associated with its deployment. In-depth comprehension of AI combined with a detailed examination of random forest prepares us for venturing into a new field in which human–machine boundary is blurred, opening the way to new perspectives and wide possibilities.

Chapter 6 is dedicated to the interaction between AI and linear regression. This symbiotic relation is a valuable asset for the AI arsenal, allowing the use of advanced techniques, such as the polynomial regression, ridge regression and LASSO to overcome the limitations of simple linear regression, while preserving its fundamental philosophy of clear interpretation of data. This chapter explores in detail the linear regression, its operation and practical application in the context of AI.

Chapter 7 is dedicated to the application of the XGBoost machine learning algorithm to a practical case study focused on the prediction of machine failures. It starts by describing the dataset used, by providing details on the key characteristics and the target variable related to machine failure indicators. The XGBoost algorithm implements predictive modeling, by refining the hyperparameters to optimize performances using cross-validation techniques. It evaluates model performances using measures such as accuracy, precision, return and ROC curve.

Chapter 8 presents an approach dedicated to the evaluation of the structural operability of intelligent systems of systems (SoS). The proposed evaluation of structural operability can be used in several ways to enhance the performance and efficient operability of the intelligent SoS. The user interface layer may provide users with operability indicators that facilitate the monitoring and understanding of the current state of the system and making informed decisions. By integrating these metrics proposed in the algorithms of analysis and intelligence of the data collection, analysis and intelligence layers, the latter can identify the potential barriers to interoperability and efficient communication between CS, thus anticipating potential problems and taking preventive measures to solve them. The evaluation of structural operability can be integrated at each layer of the architecture of intelligent SoS to provide a continuous evaluation of the system operability. This facilitates proactive decision-making and real-time adjustments that guarantee the optimal operation of the intelligent SoS, despite the potential challenges related to CS interoperability.

Finally, this book is dedicated to methods and application of AI in mechanics. Each chapter contains a clear presentation of the techniques used and developed, accompanied by illustrative examples. This book is addressed to students and is also a valuable aid to engineers and academic researchers.

Abdelkhalak EL HAMI
November 2024

1

Dynamic of PP-GF70 Using an Intelligent Method

1.1. Introduction

During a high-speed accident, a car is subjected to gradual stresses that generate localized crumpling and plastic hinges leading to significant deformations throughout its many subsystems. The structure is locally deformed when materials reach their yield stress or the critical breakage loads, which creates localized structural deformations during the wave propagation transit time, and then mass and stress effects. This is due to the transient response and inertia. Crashes are dynamic phenomena that last between 100 and 160 milliseconds (Bois et al. 2004).

In case of temporal loading, the crash phenomenon generally depends on time. If the applied force is cyclic and the frequency is below one-quarter of the natural frequency of the structure, the problem is quasi-static. Dynamic analysis is required if the load is applied more frequently or suddenly (Cook et al. 2002). The stiffness matrix for the dynamic computation is similar to that for static computation, but it also requires mass and damping matrices (El Hami and Radi 2017). The influence will be discussed in the collision analysis, where the structures are subjected to exceptionally high stresses for a short period of time. The collision phenomenon is discussed, which is a temporary response of the structure.

The response requires time integration of the differential equations of motion. The load applied to the structure produces several frequencies upon crash, and the response can only be computed for several multiples of the longest period. In this case, direct explicit integration may be appropriate. Various parameters are used to determine the impact strength of the structure depending on the nature of the impact,

---

Chapter written by Abdelilah ELBAZZE and Bouchaib RADI.

and during the construction of the entire car structure, this quality is considered as the most important. After a collision, this criterion describes the capacity of a structure to protect its passengers. The front, rear and side structures of a car have evolved in time to reduce the impact and preserve the integrity of the passenger compartment (Johnson and Mamalis 1978). The resistance to impact can be determined either in advance, using numerical models or experiments, or retrospectively, by examining accident-related data. Resistance to impact is assessed based on the examination of the structure deformation and energy absorbed during impact, the vehicle acceleration upon impact and also the risk of harm assessed using human body models.

Our study involves the use of *crashworthiness* to redimension the structure of our future cars, by integrating composite materials lighter than steel, which have a good mechanical and dynamic behavior. The work conducted in this chapter involves a comparison of the results of side impact between a steel body and a body that integrates a thermoplastic composite (PP-GF70). This material represents an innovative solution in several industries due to these mechanical and thermal advantages, and also to its manufacturing process. In the automotive industry, the race to save energy and reduce pollutant gas emissions drives an increasing interest of builders and equipment manufacturers in using low-density materials (Al-Maghribi 2008).

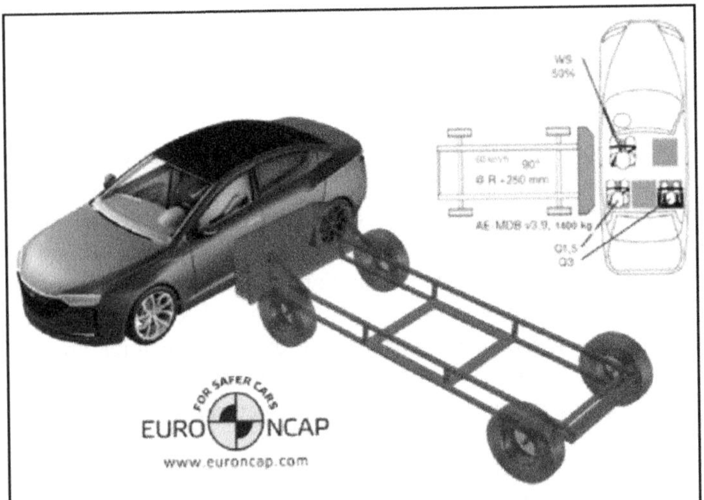

EURONCAP AEMDB Side crash regulation applied on G3 model.

Figure 1.1. *Side-crash simulation using a mobile barrier (Morancay and Winkelmuller 2009). For a color version of this figure, see www.iste.co.uk/elhami/applicationsofai.zip*

1.2. Crash analysis

The dynamic analysis uses the same stiffness matrix as the static analysis, but it also requires mass and dampening matrices. For a given loading amplitude, the dynamic response can be higher or lower than the static response. It will be much more significant for cyclical loading with a frequency near the natural frequency of the structure (ESI Group 2012).

If loading excites only some of the lowest frequencies and the response must be computed over a period of time equal with several multiples of the longest vibration period, as it is the case for seismic loading, the mode-superposition method or an implicit direct integration method may be appropriate (ESI Group 2012).

In the crash analysis, structures are subjected to extremely strong forces during the short-impact time. Here, we are looking for a transient response known as response history. The solution requires time integration of the differential equations of motion. During the crash, the loading excites many frequencies and the response has to be computed only on several multiples of the longest period. In this type of case, an explicit direct integration may be appropriate (ESI Group 2012).

1.2.1. *Dynamic equations*

1.2.1.1. *General equation*

The equation governing the structural dynamics, derived below, provides general expressions for structural mass and dampening. The equation of motion is derived by requiring that the work done by external loads is equal to the sum of work absorbed by dissipative and inertial forces for all virtual displacement. For a single element of volume V and surface S, this equilibrium of work becomes (ESI Group 2012),

$$\int\{\delta u\}^T\{F\}dV+\int\{\delta u\}^T\{\Phi\}dS+\sum_{i=1}^{n}\{\delta u\}_i^T\{P\}_i=\int\left(\{\delta u\}^T\rho\{\ddot{u}\}+\{\delta u\}^T c\{\dot{u}\}+\{\delta\varepsilon\}^T\{\sigma\}\right)dV \quad [1.1]$$

where

– $\{F\}$: body forces;

– $\{\Phi\}$: traction surface;

– $\{P\}_i$ and $\{\delta u\}_i$ are the prescribed concentrated loads at nodes and their displacements;

– $\{\delta u\}$ and $\{\delta \varepsilon\}$ are the virtual displacements and their corresponding strains.

FE discretization yields (ESI Group 2012),

$$\{u\} = [N]\{d\}; \quad \{\dot{u}\} = [N]\{\dot{d}\}; \quad \{\ddot{u}\} = [N]\{\ddot{d}\}; \quad \{\varepsilon\} = [B]\{d\} \qquad [1.2]$$

where:

– [N]: shape functions (space functions);

– $\{d\}$ is the nodal degrees of freedom (time functions).

Hence, equations [1.2] represent a local separation of variables. Combining equations [1.1] and [1.2] (ESI Group 2012) yields:

$$\{\delta d\}^T \left[\int \rho[N][N]dV\{\ddot{d}\} + \int c[N]^T[N]dV\{\dot{d}\} + \int [B]^T\{\sigma\}dV - \int [N]^T\{F\}dV \int [N]^T\{\Phi\}dS - \sum_{i=1}^{n}\{P\}_i \right] = 0 \qquad [1.3]$$

The first two integrals of equation [1.3] are identical as "coherent" elementary mass and damping matrices (ESI Group 2012):

$$[m] = \int \rho[N]^T[N]dV \qquad [c] = \int c[N]^T[N]dV \qquad [1.4]$$

The term "coherent" highlights the fact that these shapes directly derive from the EF discretization and use the same shape functions as the element force matrix. The internal force vector of the element $\{r^{int}\}$ is defined as forces and torques applied to elements with 6 d.o.f. per node to resist stresses inside the element (ESI Group 2012):

$$\{r^{\mathrm{int}}\} = \int [B]^T\{\sigma\}dV \qquad [1.5]$$

A similar notation is used to identify forces and torques applied to nodes following the application of external loads on the element (ESI Group 2012):

$$\{r^{ext}\} = \int [N]^T\{F\}dV + \int [N]^T\{\Phi\}dS + \sum_{i=1}^{n}\{P\}_i \qquad [1.6]$$

The expression in square brackets in equation [1.3] should be eliminated for the equation to be valid for an arbitrary $\{\delta d\}$. According to the notations of equations [1.4]–[1.6], equation [1.3] yields (ESI Group 2012):

$$[m]\{\ddot{d}\} + [c]\{\dot{d}\} + \{r^{int}\} = \{r^{ext}\} \qquad [1.7]$$

Equations [1.5] and [1.7] are valid for the properties of linear and nonlinear materials. If the material is linear elastic, then the loads associated with the element stresses are $\{r^{int}\} = [K]\{d\}$, where [k] is the classic element strength matrix. The global shapes for a structure with multiple elements are:

$$[M]\{\ddot{D}\} + [C]\{\dot{D}\} + \{R^{int}\} = \{R^{ext}\} \quad [1.8]$$

$$[M]\{\ddot{D}\} + [C]\{\dot{D}\} + [K]\{D\} = \{R^{ext}\} \quad [1.9]$$

1.2.1.2. *Direct integration methods*

The determination of the response history using a step-by-step integration of dynamic equations is known as direct integration. The response is evaluated at instants separated by time intervals Δt. Therefore, we calculate the structural displacements at instants Δt, 2 Δt, 3 Δt, …, n Δt, etc. At n time steps, the equation of motion (equation [1.7] or equation [1.9]) is:

$$[M]\{\ddot{D}\}_n + [C]\{\dot{D}\}_n + \{R^{int}\}_n = \{R^{ext}\}_n \quad [1.10a]$$

or

$$[M]\{\ddot{D}\}_n + [C]\{\dot{D}\}_n + [K]\{D\}_n = \{R^{ext}\}_n \quad [1.10b]$$

Time discretization is made using finite difference approximations of time derivatives. Direct integration methods calculate the conditions at the time step n + 1 from the equation of motion, with a difference expression and the conditions known at one or several previous time steps. Algorithms can be classified as explicit or implicit. An explicit algorithm uses a difference expression of the general form that contains only historical information on its right-hand side

$$\{D\}_{n+1} = f\left(\{D\}_n, \{\dot{D}\}_n, \{\ddot{D}\}_n, \{D\}_{n-1}, \ldots\right) \quad [1.11]$$

The difference expression is combined with the equation of motion, equation [1.10] at the time step n. An implicit algorithm uses an expression that differs from the general form (ESI Group 2012):

$$\{D\}_{n+1} = f\left(\{\dot{D}\}_{n+1}, \{\ddot{D}\}_{n+1}, \{D\}_n, \{\dot{D}\}_n, \{\ddot{D}\}_n, \ldots\right) \quad [1.12]$$

which is combined with the equation of motion at the time step n + 1.

1.2.1.3. *Explicit analysis and implicit analysis*

In the practical application, significant differences between explicit and implicit methods are related to stability and economy. Explicit methods are conditionally stable, which means there is critical time step Δt_{cr} that should not be exceeded, otherwise, the numerical process could "explode" and become unstable. As Δt_{cr} is quite small, many time steps are required, but each one is rapidly executed. Implicit methods are unconditionally stable, which means that calculations remain stable irrespective of the size of Δt (although accuracy is impacted). In explicit methods, the matrix of coefficients of $\{D\}_{n+1}$ can be rendered diagonal, so that $\{D\}_{n+1}$ is low-cost calculated at each time step. In implicit methods, the matrix of coefficients of $\{D\}_{n+1}$ cannot be rendered diagonal, so that the cost per time step is higher, increasingly higher as the size of the FE meshing increases. Another consequence of the difference in matrix topology is that an implicit method requires more information storage space than an explicit method.

For computers, matrix multiplication is not a difficult task. Matrix inversion is a much more computationally intensive operation. The implicit code finds a solution by calculating the stiffness matrix element and then inverts it to solve the displacements in the element.

Implicit direct time integration is adapted to structural dynamics problems where loads vary slowly, similar to the loads generated by an earthquake. The response is dominated by lower modes.

This method competes with the modal method, and can be less expensive where many modes would be required in the nodal method and when the analysis does not need to extend over a long period of time.

Implicit methods are used because many problems encountered in practice are difficult, and for them, the use of an explicit method requires time steps that are too small for the result error to remain limited (see numerical stability). For such problems, to reach a given accuracy, much less computing time is required by an implicit method with more significant time steps.

The explicit direct integration is adapted to wave propagation problems, for example, the effects created by the explosion or by the impact load, as in the analysis of vehicle resistance to impact. High-frequency modes must be represented in the analysis. Nonlinearity can be adapted with relative ease (ESI Group 2012).

In conclusion, an implicit analysis is similar to the explicit analysis, adding after each increment implicit Newton–Raphson iterations to impose the equilibrium of forces of the internal structure with the loads applied from outside. A drawback is that during Newton–Raphson iterations, the stiffness matrix must be updated and rebuilt for each iteration. This can be computationally intensive. The equilibrium is generally applied to a certain tolerance specified by the user. Though unconditionally stable, the implicit methods generally require many time steps in order to track the physical phenomena being studied.

The explicit analysis generally requires a small time step. This leads to a large number of time steps to be calculated, which is not a problem, as the time step processing cost is low. Data storage needs are also low. The implicit analysis requires a matrix inversion. The solution to nonlinear equations requires iterative solution strategies. On the other hand, the explicit analysis does not require any iteration and matrix inversion.

The implicit methods are well adapted to solve static problems. The explicit methods are well adapted to solve dynamic problems.

1.2.2. *Explicit simulation*

The explicit analysis approaches the structure as a dynamic problem and solves the dynamic equations of motion in the time domain. It is particularly efficient to solve problems of crash, impact and similar dynamics, particularly in case of nonlinearity (plasticity, etc.) of the material, large deformations or contact. For the explicit analysis, all of the nodal and elementary quantities are given as a function of time (Zienkiewicz 2005). An explicit analysis is applied to large bandwidth excitations. RADIOSS© crash simulation is achieved with an explicit time integration using the half-step central difference scheme (Morancay and Winkelmuller 2009).

1.2.2.1. *Explicit direct integration*

As illustrated in Figure 1.2, the scheme of the half-step central difference method is: with Δt the time step; the velocity and acceleration at the time step n are approximated by central difference equations (ESI Group 2012):

$$\{\dot{D}\}_{n-1/2} = \frac{1}{\Delta t}\left(\{D\}_n - \{D\}_{n-1}\right) \quad [1.13a]$$

and

$$\{\dot{D}\}_{n+1/2} = \frac{1}{\Delta t}\left(\{D\}_{n+1} - \{D\}_n\right) \quad [1.13b]$$

$$\{\ddot{D}\}_n = \frac{1}{\Delta t}\left(\{\dot{D}\}_{n+1/2} - \{\dot{D}\}_{n-1/2}\right) = \frac{1}{\Delta t^2}\left(\{D\}_{n+1} - 2\{D\}_n + \{D\}_{n-1}\right) \quad [1.14]$$

The equation of motion is rewritten, equation [1.10] with a velocity shift by half-time step. Therefore for the half-step method, we use the following equations (ESI Group 2012):

$$\{D\}_{n+1} = \{D\}_n + \Delta t\{\dot{D}\}_{n+1/2} \quad [1.15]$$

$$\{\dot{D}\}_{n+1/2} = \{\dot{D}\}_{n-1/2} + \Delta t\{\ddot{D}\}_n \quad [1.16]$$

$$[M]\{\ddot{D}\}_n + [C]\{\dot{D}\}_{n-1/2} + \{R^{int}\}_n = \{R^{ext}\}_n \quad [1.17]$$

These three equations combined yield:

$$\frac{1}{\Delta t^2}[M][D]_{n+1} = \{R^{ext}\}_n - \{R^{int}\}_n + \frac{1}{\Delta t^2}[M]\left(\{D\}_n + \Delta t\{\dot{D}\}_{n-1/2}\right) - [C]\{\dot{D}\}_{n-1/2}$$

[1.18]

where, if the linear conditions prevail, $\{R^{int}\}_n = [K][D]_n$.

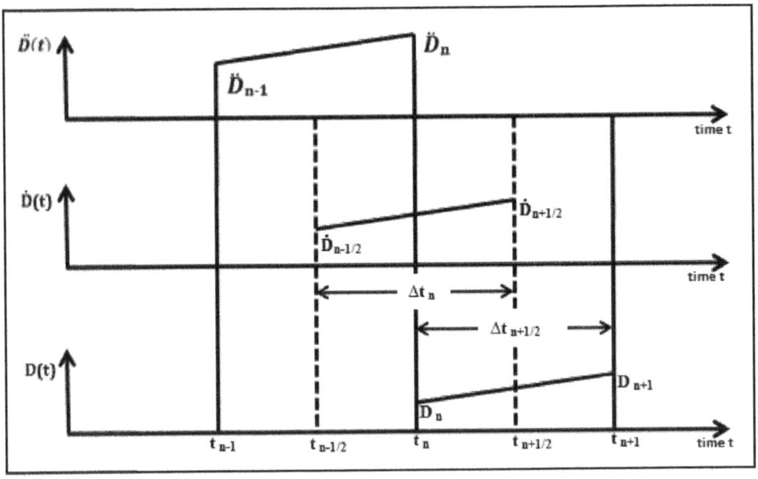

Figure 1.2. *Explicit time integration (ESI Group 2012)*

1.2.2.2. *Numerical stability*

The central difference scheme is conditionally stable, so that the time step Δt is limited. The shell time step required for stability is calculated with the PAM-CRASH software package (Morancay and Winkelmuller 2009) as follows:

– $\Delta t_{\text{shell}} = \text{kMIN}\left[\text{L} / \text{c} \min\left(1.0, \frac{\text{L}}{\sqrt{3\text{h}}}\right)\right]$ for active bending;

– $\Delta t_{\text{shell}} = \text{kMIN}\left[\text{L} / \text{c}\right]$ when bending is not active.

where k is the scale factor for the time step, MIN is the minimum on all the shell elements, L is the characteristic length and c is the sound velocity:

– $c = \sqrt{E / \rho}$ for a nonactivated Poisson ratio;

– $c = \sqrt{E / \rho\left(1 - \nu^2\right)}$ for an activated Poisson ratio;

where h is the thickness of the shell, ρ is the density and E is Young's modulus.

NOTE.– If a damping ratio ξ (KSI) was specified among the properties of the material, Δt_{shell}, the stable time step of the shell is reduced by a factor $\left(\sqrt{1 + \xi^2} - \xi\right)$ (Morancay and Winkelmuller 2009).

1.2.2.3. *Calculation of the time step*

Generally, the smallest time step of all of the elements of a model is used for time integration. For 2D (shell) elements and 3D (solid) elements, the calculation of the time step relies on a "characteristic" length. There are two time step criteria available in PAM-CRASH (Morancay and Winkelmuller 2009):

– Standard time step criteria (LARGE): standard time step criteria for quadrilateral, degenerate quadrilateral, triangular and C° triangular shell are based on the travel time of the elastic wave along the characteristic lengths illustrated in Figure 3.3. The use of standard time step criteria is generally reflected by time steps greater than those resulting from strict criteria (Morancay and Winkelmuller 2009).

– Strict time step criteria (SMALL): strict time step criteria for the shells rely on the travel time of the elastic wave along the characteristic lengths illustrated in Figure 1.3 (Morancay and Winkelmuller 2009).

Strict time step criteria should ensure the stability of the solution in serious cases (highly strained meshes), where standard economic criteria may become inadequate (Morancay and Winkelmuller 2009). The use of rigorous time step criteria may sometimes significantly reduce the time step solution compared to the time step solution based on standard criteria (Morancay and Winkelmuller 2009).

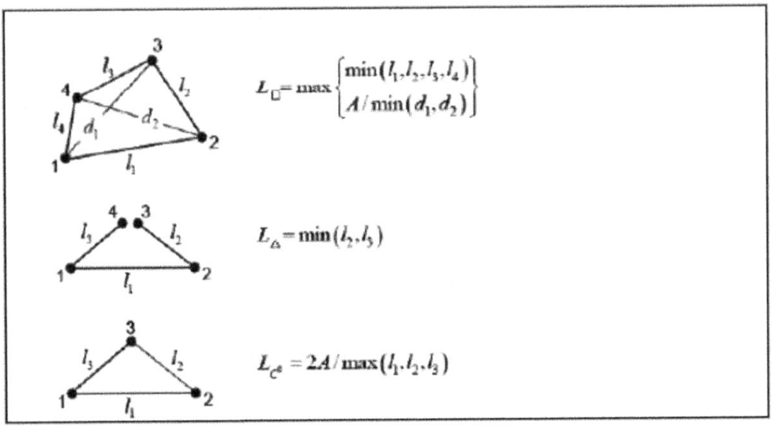

Figure 1.3. *Time step of the shell element (Morancay and Winkelmuller 2009)*

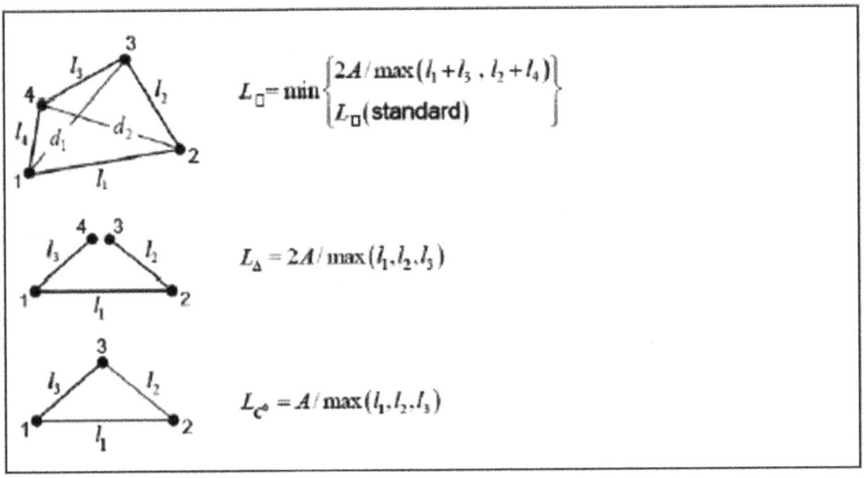

Figure 1.4. *Rigorous time step of the shell element (Morancay and Winkelmuller 2009)*

1.2.3. *Shell elements*

1.2.3.1. *Reduced and selective integration*

Upon the evaluation of stiffness matrices of elements, a Gaussian quadrature rule is commonly used. Nevertheless, a complete quadrature rule may lead to erroneous results in some problems. It is the case in the presence of thin plate and shell elements, where the shear terms may drive a solution that is too rigid for bending problems, generally known as "locking". A common remedy relies on reduced integration techniques, in which a smaller number of Gauss integration points is used to evaluate these terms. Normally, selective reduced integration or reduced integration techniques can be used to avoid shear "blocking" in thin plates and shells (Cook et al. 2002). In selective reduced integration, the bending term is integrated to the normal rule, while the shear term is integrated to a lower order rule. In reduced integration, the bending and shear terms of the stiffness matrices of elements are integrated to the same reduced rule (Cook et al. 2002). A significant gain in computation time can also lead to serious problems, but special stabilization terms must be added in order to avoid parasitic strain models, associated with zero energy deformation modes, which may be induced in the solution (El Hami and Radi 2017).

1.2.3.2. *Hourglass phenomenon*

As said by Al-Maghribi (2008): "hourglass modes are nonphysical, zero-energy modes of deformation that produce zero strain and no stress. Hourglass modes occur only in under-integrated (single integration point) solid, shell and thick shell elements…". The hourglass problem is very common for under-integration techniques used in the numerical integration of stiffness (or other quantities) and it is particularly common in an element of integration of points used in explicit codes (Zienkiewicz 2005). It is a phenomenon that can be easily excited in regular meshed structures, or a poor loading (Zienkiewicz 2005). Hourglass modes are essentially some modes of element deformation that erroneously predict zero strain energy at the points of element integration (Zienkiewicz 2005). The following methods could be used to control or eliminate the hourglass problem (Zienkiewicz 2005):

– Fully stress all the nodes at the boundary (this would be the best modeling approach).

– Use a different type of element that has no hourglass mode, for example, "fully" integrated or "selectively reduced" elements (drawbacks: more expensive and more unstable in large deformation applications such as the collision simulation), or triangular elements (concerning its too rigid behavior in some applications).

– Refine the mesh.

1.3. Application to automobile crashes

The case study involves making the frontal impact scenario in the crash simulation, the objective being to compare the results obtained for steel and PP-GF70 taking into account the breakage in this material by the finite element method using RADIOSS as solver, ANSA as preprocessing software and Meta-post as post-processing software (see Figure 1.5).

This study has a more in-depth approach of the dynamic behavior of the structure compared to the two previous studies. In this chapter, the objective of the work is to compare the global behavior of the vehicle during the crash by testing several concepts of PP-GF70-targeted parts.

Figure 1.5. *Example of a car submitted to the side-crash test. For a color version of this figure, see www.iste.co.uk/elhami/applicationsofai.zip*

1.3.1. *Description of the study*

The next section of our study focuses on the simulation of the scenario of vehicle side crash with a mobile barrier, the objective being to conduct two simulation tests: one without modifying the materials of the car body, and the other by injecting our study material in the modeled car body. The simulation involves several phases of preparation before starting the computation with Radioss solver:

– Model preparation according to ANSA: collect all of the CAD data for the vehicle in batch form, especially the components concerning the side-crash simulation, then conduct the generation of neutral fibers for sheet metal parts. Once these two stages have been completed, the next one is the generation of meshing of all of the parts and the assembly of each subsystem of the vehicle. Then, thicknesses, nuances and also the parameters of the properties applied to the parts are verified, by comparing them with CAD. Once this stage is validated, subsystems are reassembled with the body via rigid connections that will be described in detail in the next sections. Then, the last stages are: interfacing, model instrumentation and configuration of the scenario by setting all of the conditions required to achieve it (barrier integration, applied velocity, results output, etc.).

– Launching computation with Radioss: once the model is ready, input files are generated in order to create the crash scenario in the solver.

– Post-processing of results under Meta-Post: in general, we verify the global behavior of the scenario, and then the area targeted by our study (absorbed energy, displacements, plastic deformation).

1.3.2. *Model preprocessing*

In the following sections, a description of the different steps for constructing an FE model of the vehicle will be provided, with the aim of analyzing the crash simulation results.

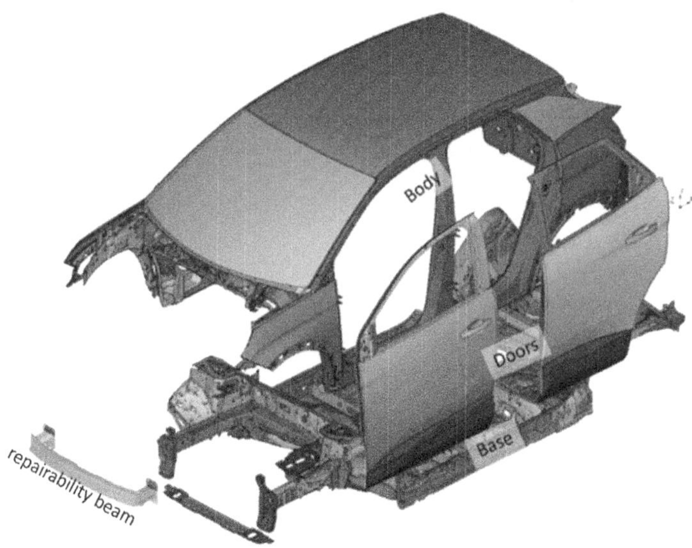

Figure 1.6. *List of vehicle components. For a color version of this figure, see www.iste.co.uk/elhami/applicationsofai.zip*

1.3.3. *Mesh generation according to ANSA*

After having run the ANSA software, the CAD file that contains the geometry of all of the components of the car is loaded. The geometry of the structure will be available on the main screen, as shown in Figure 1.7. The CAD is used to generate the automatic meshing of all of the parts that serve to build the subsystems of the vehicle. These are then assembled.

Figure 1.7. *Meshing of the car parts. For a color version of this figure, see www.iste.co.uk/elhami/applicationsofai.zip*

1.3.4. *Materials and properties*

The use of various types of steel when building a car body is represented in Figure 1.8. The highest percentage of all of the steels used in car building is that of dual-phase steel.

During the crash simulation stage, the modeling of materials is essential for correlating the results of simulation with the trial for a good prediction.

As a general rule, each car manufacturer has a numerical database on all of the materials, which it uses in numerical computation, so in crash modeling most materials are numerically modeled, based on the Johnson–Cook model.

(a)

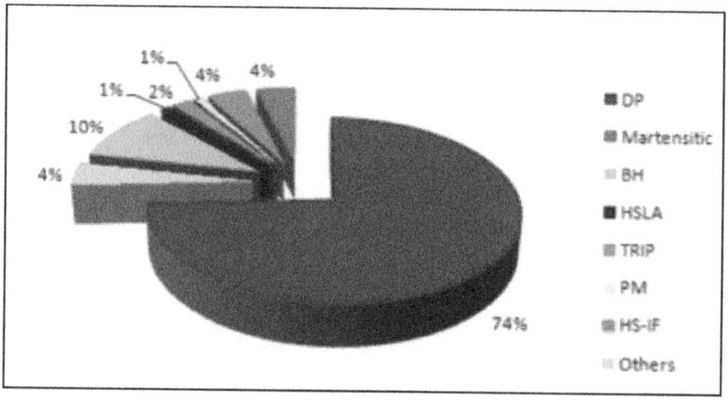

(b)

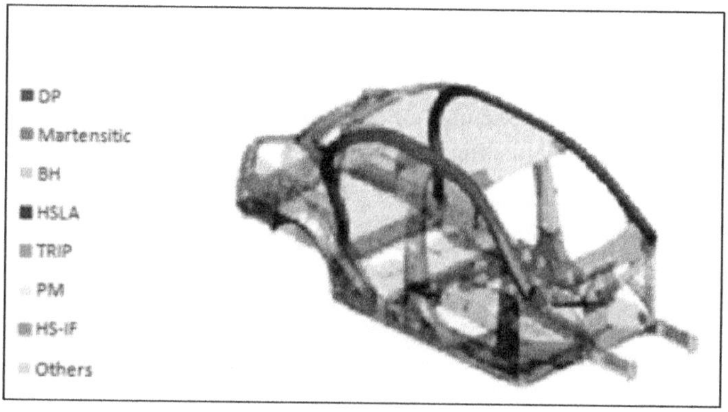

Figure 1.8. *Distribution of materials in the car body (Al-Maghribi 2008). For a color version of this figure, see www.iste.co.uk/elhami/applicationsofai.zip*

1.3.5. *Generation of part assembly connections*

The procedure starts with the insertion of connection files created by the design department in charge with the car. These files contain all of the various types of connection that were defined for the assembly of parts in each component.

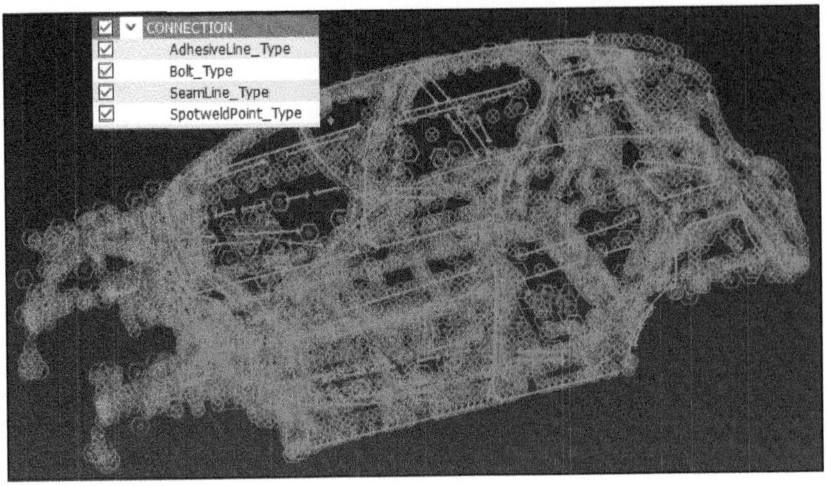

Figure 1.9. *Various types of connection in the car. For a color version of this figure, see www.iste.co.uk/elhami/applicationsofai.zip*

1.3.6. *Assembly of components*

The next stage after the assembly of parts of each car component is the setup of these components in the main structure of the car (body).

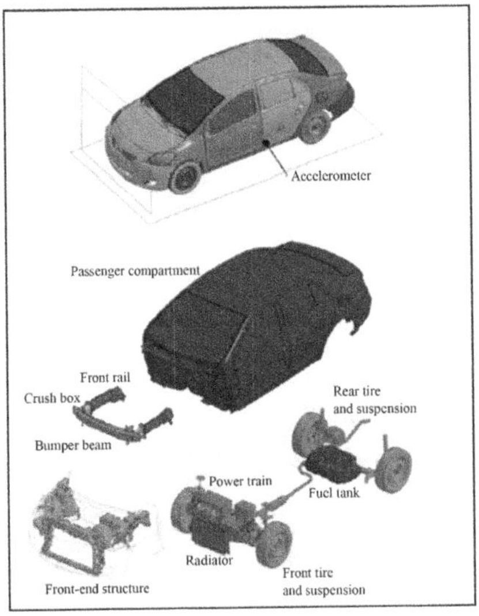

Figure 1.10. *Various subsystems assembled in the car body. For a color version of this figure, see www.iste.co.uk/elhami/applicationsofai.zip*

These components (underbody components, the front-end part or the rear part) are connected to the body using various types of connections: screws, rivets, circlips … or another fixing type. They are modeled as follows.

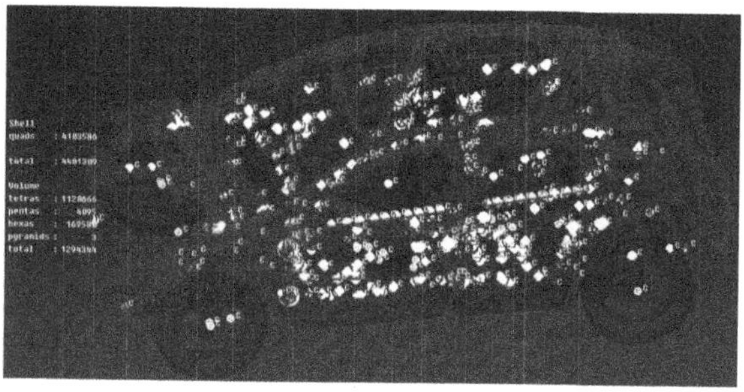

Figure 1.11. *Modeling of the connectivity between subsystems and the body. For a color version of this figure, see www.iste.co.uk/elhami/applicationsofai.zip*

Connectivity is generally a beam-type spring linked to two rigid bodies, each of which is attached to the concerned structure (body or subsystem). Once the subsystems are assembled and the connectivity is verified, our FE model completed the first preprocessing stage, which makes it possible to approach the next stage of verification of the FE model concerning the interfacing and intersections between the subsystems.

1.3.7. *Interfaces*

In the explicit computational codes, the management of interfaces and intersections is highly important, either at the start of the FE model preparation or during the computation, in order to obtain a better prediction of the results with respect to the trials. There are two ways in which the intersections in our FE model under ANSA are corrected.

In the case of small intersections related to the meshing of parts with respect to the neutral fiber, the correction is required by shifting the nodes of the elements of each part. However, in the case of large intersections between, for example, the subsystems, the problem is related to the co-evolution of the latter, in this case disinterfacing to be done during the creation of the interfaces or an adaptation to be made on one of the subsystems to avoid intersection.

Interfaces are defined in order to model how the parts interact when in contact with one another. In any interaction between surfaces or situation in which two components collide. The two terms that are commonly used are slave and master (see Mahindre 2020).

The master assigns the task to the slave or, to put it differently, the slave contacts the master and it executes the motion assigned by the master. The most commonly used method approaches the behavior of the slave nodes in contact with the master segment as a spring that generates resistance forces as a function of penetration (see Mahindre 2020). Spring elements are created between the slave nodes and its projection.

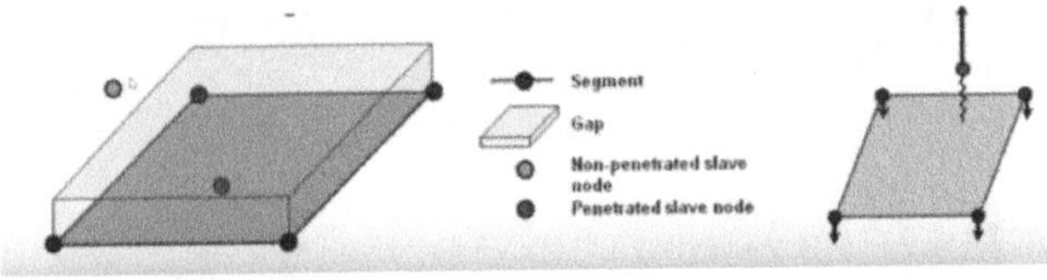

Figure 1.12. *Interface modeling in Radioss (see: Mahindre 2020). For a color version of this figure, see www.iste.co.uk/elhami/applicationsofai.zip*

Hence, when the slave node gets close to the master and it is near the imaginary master space, radios will form a spring between the slave node and this surface or particular segment.

1.3.8. *Side crash with mobile barrier*

Side-crash accidents are rated as causing the second highest frequency in deaths and severe injuries. Compared to a head-on impact, inside the car cab there is little side space to absorb the energy and serious head and chest injuries are frequent.

Figure 1.13. *Side crash with a mobile barrier (Euro NCAP 2020). For a color version of this figure, see www.iste.co.uk/elhami/applicationsofai.zip*

This test can be used to ensure the presence of an adequate protection of the critical areas of the body. This led to the reinforcement of peripheral structures of pillar B (between front and rear doors) of vehicles, to the installation of side airbags or curtain airbags, and also to the development of less visible energy absorbing structures integrated in the seats and the panels of the car doors (Euro NCAP 2020).

The objective of this study is to reinforce the car structure by working on the rear zone of the sub-frame, with the purpose of protecting the rear-seat passengers and also of lightening the car body by integrating composites in the car structure, which is assumed to absorb the impact energy.

The numerical computation was conducted on this type of crash in order to verify the results of the impact between steel, as standard material for the car body, compared with the thermoplastic composites by choosing polypropylene reinforced with long glass fibers with 70% of the total volume. This material is applied in a structural part of the body that will be strongly stressed in the case of this type of impact.

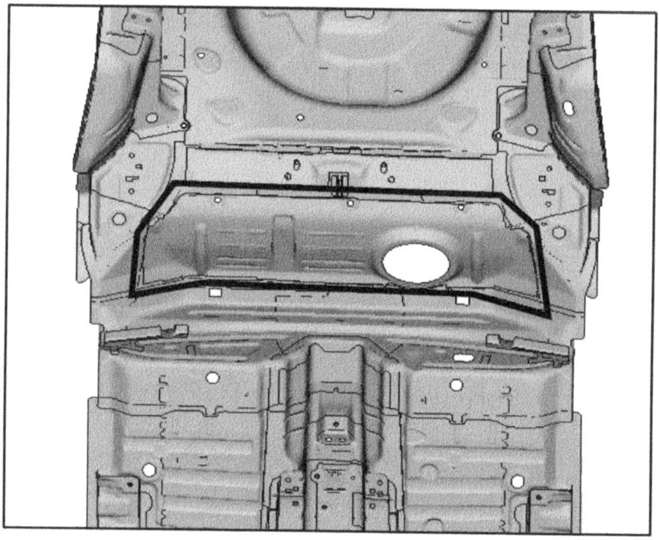

Figure 1.14. *Target part to be studied in the accident*

It represents a large part of the structure where the back seat is fixed. The geometric characteristics of the part made from steel and from PP-GF70 are represented in Figure 1.15.

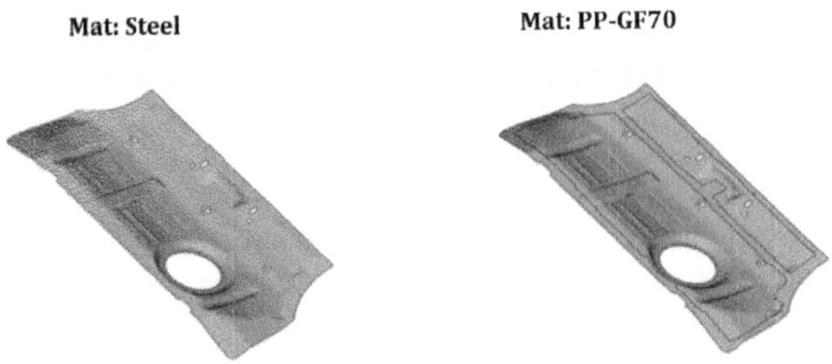

Figure 1.15. *Shape of the target part and thickness distribution for each material. For a color version of this figure, see www.iste.co.uk/elhami/applicationsofai.zip*

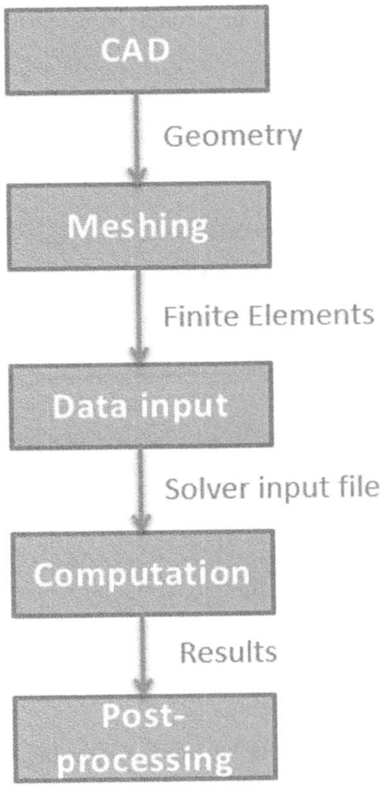

Figure 1.16. *Descriptive diagram of the various steps to simulating a crash scenario. For a color version of this figure, see www.iste.co.uk/elhami/applicationsofai.zip*

This comparative study uses the same geometric shape of the part for the two materials, but the thickness values of the composite material part were modified (ranging between 1 mm and 4.5 mm depending on the impact zone), while steel thickness is homogeneous (1 mm) throughout the part.

After all of these stages of model preparation aimed at comparing the dynamic responses of our material with respect to steel, Figure 1.16 describes the various stages to achieve our crash scenario simulation.

1.4. Numerical results and discussions

1.4.1. *Numerical stability of the model*

1.4.1.1. *Energy balance*

The stability of energy curves during the computation is an essential criterion for the validation of the numerical state of the model. In other terms, the total energy should remain constant throughout the computation, with a variation of the kinetic energy that is converted into internal energy (Prem et al. 2015).

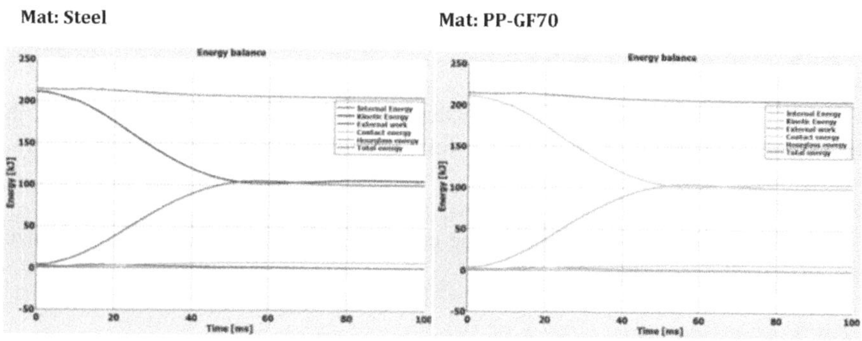

Figure 1.17. *Energy balance of a side crash. For a color version of this figure, see www.iste.co.uk/elhami/applicationsofai.zip*

Once this criterion is verified, the next step is the verification of other artificial energies, which allow the numerical stability of the model, as the hourglass energy, which makes it possible to avoid the deformation of shell and solid elements under zero force (Marur 2008), and also the contact energy, which represents the force used during computation to avoid penetrations (Muñoz 2015). These two energies remain in general very small compared to the total energy. For certain crashes, the users limit them between 5% and 10% of the total energy (Prus et al. 2017).

1.4.1.2. *Time step stability*

Time step stability in the crash is generally an essential criterion for maintaining the computation convergence, especially as the integration diagram employed strongly depends on this parameter for the stability (Bobineau and Morançay 2013.

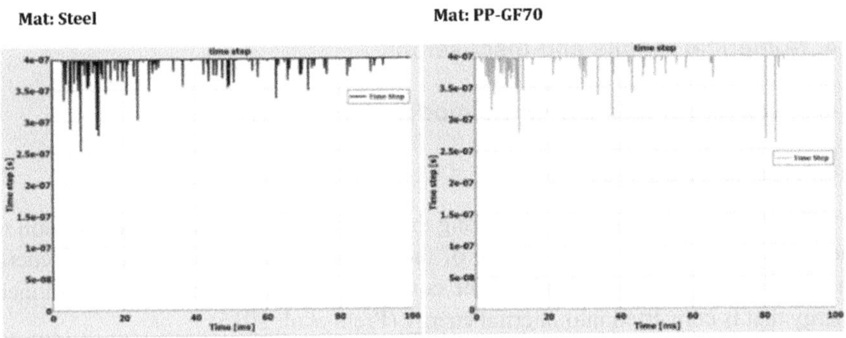

Figure 1.18. *Time step evolution during computation*

At the beginning, the time step value drops instantaneously because of the significant deformation of shell and solid elements compared to the critical value (4e-7 s).

On the other hand, at the end of the computation the time step value stabilizes around the critical value.

1.4.1.3. *Mass stability*

The evolution of the model mass during computation is due to the time step drop (Radi and El Hami 2016). This mass addition must be limited with respect to the initial mass of the model (Morancay and Winkelmuller 2009).

In the automotive industry, these values are generally limited between 1% and 2%; otherwise, the meshing of the zone where the time step drops must be reviewed or the time step value must be readjusted.

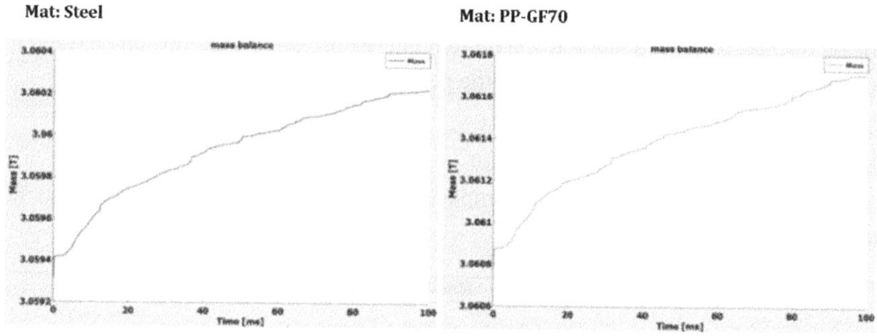

Figure 1.19. *Mass evolution during simulation*

In our model, the mass addition during the computation is about 0.8 kg, which is a very low value compared to the initial state of the computation.

1.4.2. *Model behavior*

1.4.2.1. *Acceleration of the vehicle*

The criteria for car behavior verification and validation include the measurement of acceleration during impact (see Figure 1.20).

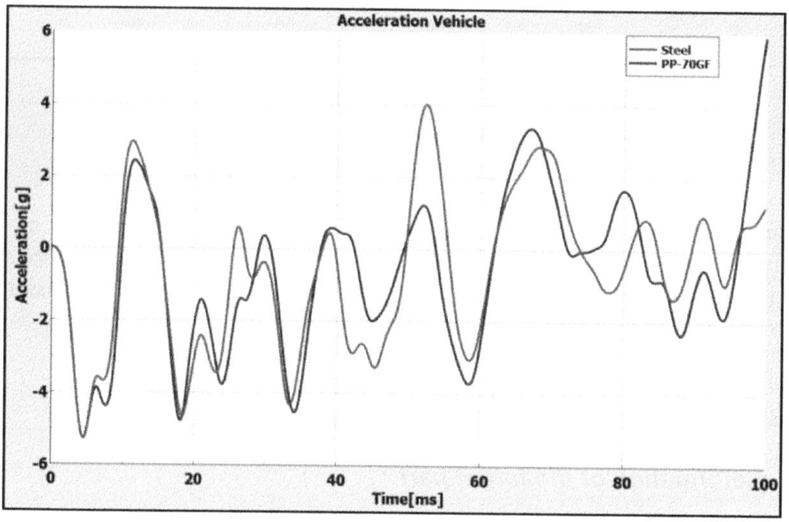

Figure 1.20. *Acceleration of the vehicle during impact. For a color version of this figure, see www.iste.co.uk/elhami/applicationsofai.zip*

In our study, the maximal value of the acceleration is about 4 g, knowing that the velocity of the car is zero at t = 0 s. The two curves are generally similar, having low amplitude values that pose no risk for the passengers.

1.4.2.2. *Absorbed energy*

The energy absorbed during the impact remains stable for the two models (see Figure 1.21).

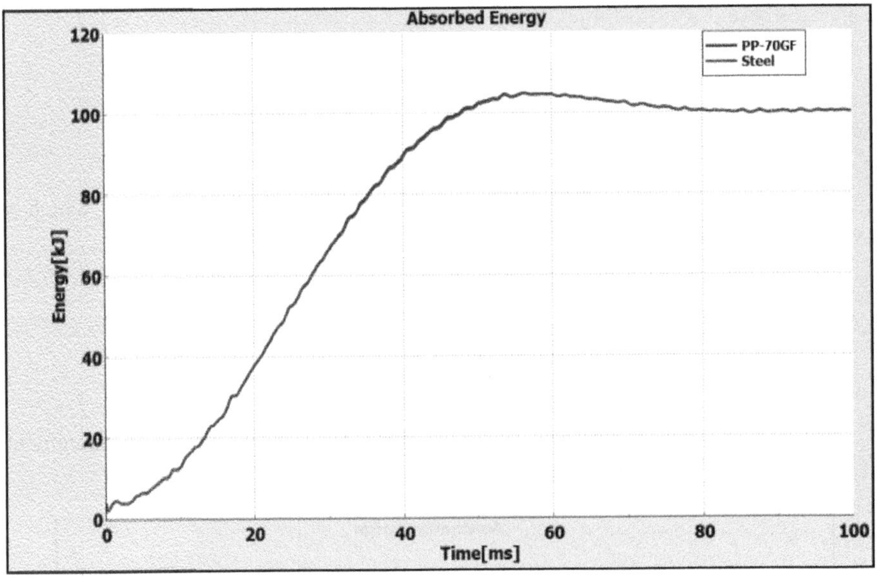

Figure 1.21. *Internal energy of the vehicle. For a color version of this figure, see www.iste.co.uk/elhami/applicationsofai.zip*

Considering the above, it can be stated that the model built with composite materials absorbs the same impact energy as the one built with steel. This result leads to the conclusion that composite materials are able to protect the passengers and reduce the transferred energy.

1.5. Behavior of the target part during impact

1.5.1. *Deformation of the target part*

The steel part starts to fold along the crash direction with a small deformation in the zone opposite to the impact (see Figure 1.22). On the other hand, the impact conducted with the PP-GF70 part does not present any deformation. It is important to note that the environment around the target part is similar for the two crashes.

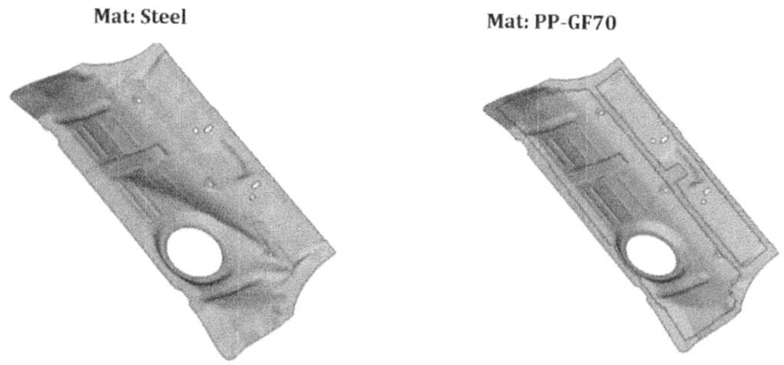

Figure 1.22. *Evolution of the deformation of the target part at the beginning of the impact. For a color version of this figure, see www.iste.co.uk/elhami/applicationsofai.zip*

The deformation continues on the steel side, with a collapse of the target part in the impact zone and the propagation of the curvature towards the zone opposite to the impact. Unlike the case of the composite, the part represents the stability with a slight deformation in the impact zone, before the elastic springback of the vehicle (see Figure 1.23).

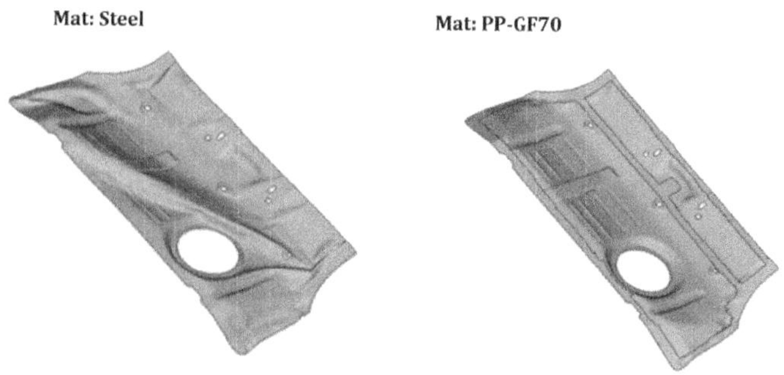

Figure 1.23. *Evolution of the deformation of the target part at maximal dynamic impact. For a color version of this figure, see www.iste.co.uk/elhami/applicationsofai.zip*

After the elastic springback of the vehicle, steel was strongly deformed compared to PP-GF70. The energy absorbed by steel is therefore higher than that of the composite. On the other hand, the risk here is having less protection for rear occupants in the case of steel.

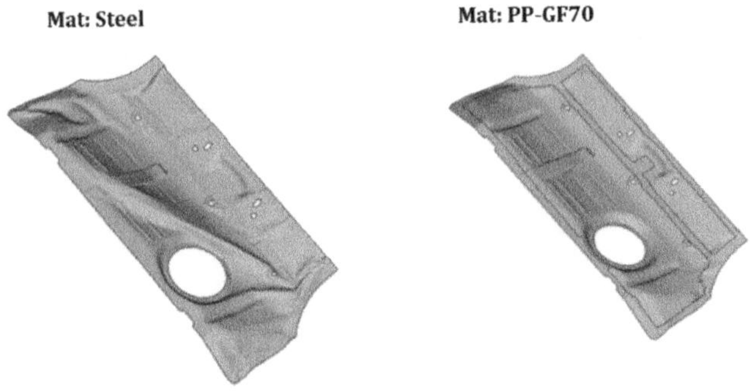

Figure 1.24. *Evolution of the deformation of the target part at the end of the impact. For a color version of this figure, see www.iste.co.uk/elhami/applicationsofai.zip*

1.5.2. *Plastic deformation*

The plastic limit authorized before steel breakage is above the plasticity generated during computation. According to the scale of iso-values, the plasticity is about 22% concentrated in the zone with respect to the impact and also in the collapse line of the part (see Figure 1.25).

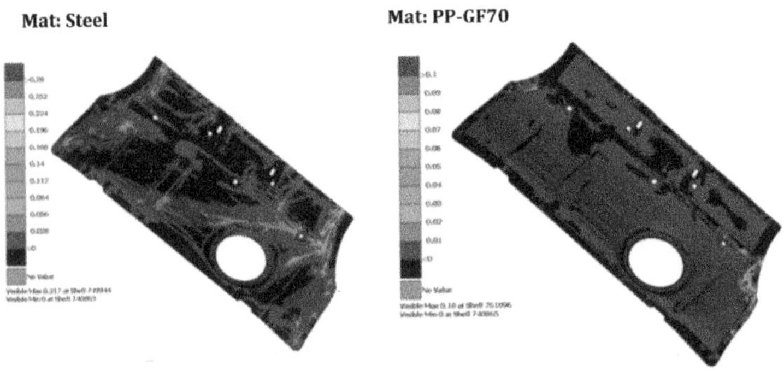

Figure 1.25. *Evolution of the plasticity of the target part at the end of the impact. For a color version of this figure, see www.iste.co.uk/elhami/applicationsofai.zip*

On the other hand, in the case of PP-GF70, the plasticity of the part exceeds the limit authorized before breakage (18% compared to the limit of 10%) concentrated in the impact zone.

1.6. Conclusion

Based on the principle of virtual work as theoretical approach – which was reinforced by a numerical approach of the finite element method and by an explicit scheme integration – this makes it possible to have a good representativity of the crash phenomenon, knowing that it is necessary to take into account the stability of the model during computation with respect to the time step and mass addition. These criteria have been verified in our case study. The energy curves revealed to be stable according to the fundamental law of dynamics with a low percentage in terms of hourglass and contact energies.

Our numerical results show that PP-GF70 is an innovative solution in the automotive industry, thanks to its performances in crash computation compared to those of steel. These results open the path for integrating this material in the car body. By reducing its weight, it contributes to the reduction of CO_2 emissions and also to savings in manufacturing costs.

1.7. References

Abo Al-Kheer, A., Eid, M., Aoues, Y., El Hami, A., Kharmanda, M.G., Mouazen, A.M. (2011). Theoretical analysis of the spatial variability in tillage forces for fatigue analysis of tillage machines. *Journal of Terramechanics*, 48(4), 285–295.

Akrout, S.E., Mhaiti, N., Radouani, M., El Fahime, B. (2021). Modeling and simulation of a vehicle crash test. *Advances in Science, Technology and Engineering Systems Journal*, 6, 937–946.

Al-Maghribi, A. (2008). Comportement des matériaux composites à fibres courtes : application à l'impact basse vitesse. PhD Thesis, Université de Toulouse.

Ben Abdessalem, A. and El Hami, A. (2015). A probabilistic approach for optimising hydroformed structures using local surrogate models to control failures. *International Journal of Mechanical Sciences*, 96, 143–162.

Ben Abdessalem, A., Pagnacco, E., El Hami, A. (2013). Increasing the stability of T-shape tube hydroforming process under stochastic framework. *International Journal of Advanced Manufacturing Technology*, 69(5–8), 1343–1357.

Bobineau, J.P. and Morançay, L. (2013). AMS and multi-domain method in RADIOSS. In *11e colloque national en calcul des structures*. CSMA, Giens.

Bois, P.D., Chou, C., Fileta, B., Khalil, T., King, A., Mahmood, H., Mertz, H., Wismans, J. (2004). *Vehicle Crashworthiness and Occupant Protection*. Automotive Applications Committee, American Iron and Steel Institute, Southfield, Michigan.

Bouyekhf, R., El Moudni, A., El Hami, A., Zerhouni, N., Ferney, M. (1996). Reduced order modelling of two-time-scale discrete non-linear systems. *Journal of the Franklin Institute*, 333(4), 499–512.

Bouyekhf, R., El Hami, A., El Moudni, A. (2001). Optimal control of a particular class of singularly perturbed nonlinear discrete-time systems. *IEEE Transactions on Automatic Control*, 46(7), 1097–1101.

Cook, R.D., Malkus, D.S., Plesha, M.E., Witt, R.J. (2002). *Concepts and Applications of Finite Element Analysis*, 4th edition. John Wiley & Sons, New York.

El Hami, A. and Radi, B. (1996). Some decomposition methods in the analysis of repetitive structures. *Computers and Structures*, 58(5), 973–980.

El Hami, A. and Radi, B. (2017). *Dynamics of Large Structures and Inverse Problems*. ISTE Ltd, London, and John Wiley & Sons, New York.

El Hami, A., Lallement, G., Minotti, P., Cogan, S. (1993). Methods that combine finite group theory with component mode synthesis in the analysis of repetitive structures. *Computers and Structures*, 48(6), 975–982.

ESI Group (2012). Virtual performance solution, solver reference manual, volume 1. General Simulation Procedure. ESI Group, Paris.

Euro NCAP (2020). Side Mobile Barrier [Online]. Available at: https://www.euroncap.com/en/car-safety/the-ratings-explained/adult-occupant-protection/lateral-impact/side-mobile-barrier.

Johnson, W. and Mamalis, A.G. (1978). *Crashworthiness of Vehicles*. Mechanical Engineering Publications Ltd, London.

Kharmanda, G., Ibrahim, M.-H., Abo Al-kheer, A., Guerin, F., El Hami, A. (2014). Reliability-based design optimization of shank chisel plough using optimum safety factor strategy. *Computers and Electronics in Agriculture*, 109, 162–171.

Kocanda, A. and Sadłowska, H. (2008). Automotive component development by means of hydroforming. *Archives of Civil and Mechanical Engineering*, 8(3), 55–72.

Mabrouk, I.B. and El Hami, A. (2019). Effect of number of blades on the dynamic behavior of a Darrieus turbine geared transmission system. *Mechanical Systems and Signal Processing*, 121, 562–578.

Mahindre, Y. (2020). To study the different types of interfaces and to compare the results of radioss simulation for each interface case [Online]. Available at: https://skill-lync.com/student-projects/to-study-the-different-types-of-interfaces-and-to-compare-the-results-of-radioss-simulation-for-each-interface-case.

Makhloufi, A., Aoues, Y., El Hami, A. (2016). Reliability based design optimization of wire bonding in power microelectronic devices. *Microsystem Technologies*, 22(12), 2737–2748.

Marur, P.R. and Srinivas, S. (2008). A reduced-order finite element model for the simulation of automotive side structure crash response. *International Journal of Vehicle Structures and Systems*, 13(2), 211–218.

Morancay, L. and Winkelmuller, G. (2009). Dynamic condensation and selective mass scaling in RADIOSS® Explicit. In *19ème Congrès Français de Mécanique*, 24–28 August, Marseille.

Moro, T., El Hami, A., El Moudni, A. (2002). Reliability analysis of a mechanical contact between deformable solids. *Probabilistic Engineering Mechanics*, 17(3), 227–232.

Muñoz, G.A. (2015). Crash analysis with RADIOSS. A study guide, 1st edition. Academic Program, Altair University, Troy.

Pickett, A. (2007). Tutorial 1 and 2, end loaded cantilever and shell element studies. Technical Report, ESI GmbH/Institute for Aircraft Design, Stuttgart.

Prem, A., Bastien, C., Dickison, M. (2015). Multidisciplinary design optimisation strategies for lightweight vehicle structures. In *10th European LS-DYNA Conference*, Würzburg.

Prus, C., Vinuesa, R., Schlatter, P., Tembras, E., Mestres, E., Berro Ramirez, J.P. (2017). Impact simulation and optimization of elastic fuel tanks reinforced with exoskeleton for aerospace applications. *International Journal of Crashworthiness*, 22(3), 271–293.

Radi, B. and El Hami, A. (2016). *Material Forming Processes, Simulation, Drawing, Hydroforming and Additive Manufacturing*, ISTE Ltd, London, and John Wiley & Sons, New York.

Zhao, Z., Yang, S., Suo, Y., Zhang, Z. (2021). Impact analysis of the dynamic added mass coefficients of ship collision based on the external analytical method. In *2020 Chinese Automation Congress (CAC)*. IEEE, Shanghai.

Zienkiewicz, O.C., Taylor, R.L., Zhu, J.Z. (2005). *The Finite Element Method: Its Basis and Fundamentals*, 6th edition. Elsevier, Amsterdam.

2

AI Integration into Additive Manufacturing

2.1. Introduction

Artificial intelligence (AI) is currently one of the most discussed technologies, among both scientists and in the public media. Though this field emerged in the previous century, several factors contributed to its development in recent years. The first one is the access to significant amounts of data. In the industrial field, the advent of Industry 4.0, which promotes the automation and sharing of data in the manufacturing technologies, led to an exponential increase in data availability and the recognition of their value. Another factor is the continuous improvement in computational power thanks to the development of increasingly high-performance processors and to algorithm optimization. Once these two limitations are overcome, most developments in AI focus on the quality of predictions.

Additive manufacturing (AM) is currently one of the most promising innovative technologies in the industrial field. AM, also known as 3D printing, revolutionized the way in which objects are designed and manufactured, offering significant advantages in terms of design freedom, prototyping speed and reducing manufacturing waste (Huang 2010; Ngo 2018; Qiu 2023).

The integration of AI in the industrial sector is a new and exciting frontier for innovation. Similar to how it has transformed many other sectors, AI application to manufacturing technologies, particularly to AM, promises a significant improvement of the design, manufacturing and quality control processes. From computer-aided design (CAD) to the optimization of printing parameters, including fault detection and real-time monitoring, AI can bring significant improvements to each stage of the AM process (Mabrouk 2019; Qiu 2023).

Chapter written by Adnane ZOUBEIR, Bouchaib RADI and Abdelkhalak EL HAMI.

This chapter closely examines the opportunities and challenges associated with integrating AI into AM. We explore the various potential applications, such as topological optimization, the prediction of performances of printed parts and the automated control of manufacturing processes. Additionally, we discuss ethical considerations and security implications related to the use of AI in this context (Kharmanda 2014).

This chapter explores the convergence of two dynamic fields: AI and AM, to address a critical manufacturing challenge. As AM evolves, it becomes increasingly evident that AI can play an essential role in the optimization and improvement of this manufacturing process (Qui 2023).

Our objective is to explore the possibilities offered by AI in the context of AM, highlighting the potential advantages, challenges and research paths to explore. The purpose of this examination is to provide valuable insights for researchers and industry professionals, while contributing to a deeper understanding of the synergies between AI and AM.

2.2. Additive manufacturing processes

2.2.1. *Principle of additive manufacturing*

Additive manufacturing (AM), commonly referred to as 3D printing, represents a new approach to creating three-dimensional objects. This method builds items layer by layer using digital data. This technology offers unparalleled design flexibility by adding material rather than subtracting it, which minimizes waste and enables the creation of complex geometries. It has become a pillar of innovation in various sectors, including aerospace, medicine, automotive and architecture (Huang 2012; LeGun 2015; Ngo 2018).

AM covers various technologies, such as stereolithography (SLA), laser powder bed fusion (LPBF), fused deposition modeling (FDM), selective laser sintering (SLS), binder jet manufacturing (BJM) and many others. Each of these technologies has its own advantages and limitations in terms of resolution, print speed, compatible materials and costs. The following section examines the shared features of these technologies, by highlighting the fundamentals of AM, the processes involved and the challenges associated with the integration of AI (Goodfellow 2016).

The significance of AM in modern industry cannot be overstated. This technology offers substantial advantages in terms of cost reduction, shorter production timelines, product customization and the creation of complex geometries that are impossible to achieve with traditional methods.

Previously known as rapid prototyping, AM has revolutionized standard approaches by allowing the creation of functional parts based on digital models. Initially used to produce prototypes, this technology rapidly extended to many other applications. The enhanced quality of the results allowed a closer integration with the final product itself, which raised questions about the term “rapid prototyping” and its relevance.

However, AM is not only limited to the production of prototypes. It now offers the possibility of manufacturing end-use parts without requiring conventional manufacturing processes. This evolution toward direct manufacturing is often referred to as “direct digital manufacturing” (DDM). This revolutionary approach has significantly reduced the manufacturing timeline and costs while offering unparalleled design freedom.

AM has thus evolved beyond its initial role of rapid prototyping to become a leading manufacturing method, paving the way for a new era of industrial production. DDM represents the future of the industry by closely integrating digital processes with manufacturing technologies, resulting in a radical transformation of how products are designed, developed and manufactured.

2.2.2. *Various stages of additive manufacturing*

The AM process typically involves several essential stages (see Figure 2.1), each playing a crucial role in the creation of the final component. Below is an overview of the typical stages involved in this process.

2.2.2.1. *Design (CAD) preparation*

In the context of AM, design preparation often involves the use of CAD software and 3D modeling tools. These tools are employed to create and optimize designs proposed by engineers and designers, taking into account both manufacturing constraints and functional requirements. This process demands specialized skills and can take from days to months, depending on the complexity of the part and the required level of detail.

2.2.2.2. *Preparation for 3D printing*

Once the design is completed, the file must be prepared for 3D printing. This involves using slicing software to divide the model into layers, generate supports and configure printing parameters. While these tasks are typically performed using specialized software, they still require human oversight to verify the settings and anticipate potential printing failures.

2.2.2.3. *Settings and material selection*

Once the file is prepared, the printer needs to be configured, and the appropriate printing parameters must be selected, such as extrusion temperature, print speed and filling density. These parameters should be adjusted based on the material used and the characteristics of the part. Optimizing these parameters can be time-consuming and requires extensive modeling and simulation to predict the material's behavior during printing. Advanced simulation tools can analyze material melting, cooling, solidification and any residual stresses that may develop. After these parameters are optimized, the printing process can begin, but it still requires continuous monitoring to detect and address any potential print issues.

2.2.2.4. *Printing*

The actual printing may be relatively quick compared to the previous stages, but it still requires constant monitoring to ensure the process proceeds smoothly. Issues such as extruder jams, printer malfunctions and quality defects may arise and require human intervention to resolve.

2.2.2.5. *Post-processing*

Once the printing is complete, the part generally requires post-processing to remove supports, smooth surfaces and perform other finishing operations. This stage also varies depending on the complexity of the part and the material used.

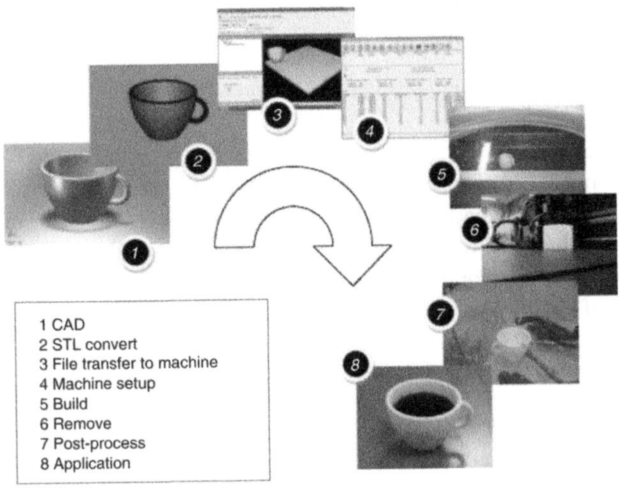

Figure 2.1. *Additive manufacturing stages. For a color version of this figure, see www.iste.co.uk/elhami/applicationsofai.zip*

Figure 2.2 shows a detailed diagram of the AM process, which concisely illustrates the various stages involved in AM, highlighting the workflow and the interconnections between different stages.

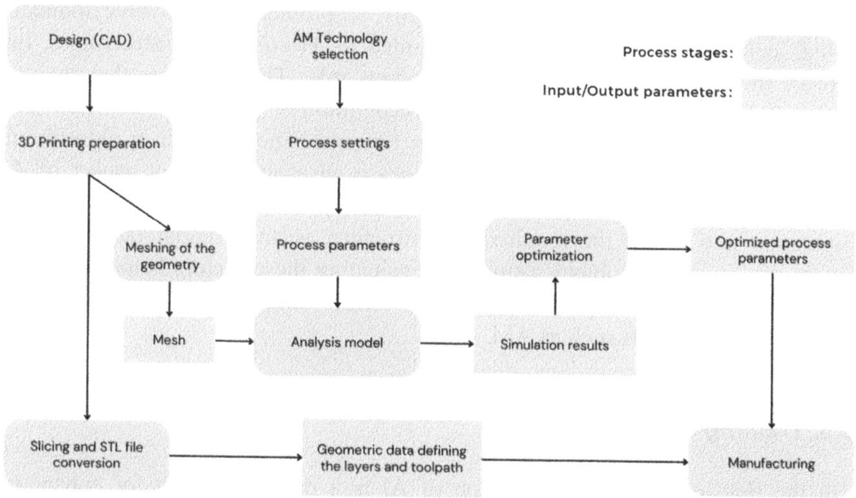

Figure 2.2. *Diagram of the additive manufacturing (AM) process. For a color version of this figure, see www.iste.co.uk/elhami/applicationsofai.zip*

This analysis of the current state of the art of AM highlights the challenges and inefficiencies that can be addressed thanks to the integration of Al. The next section explores the manners in which AI can transform these stages to improve efficiency, save time and optimize the quality of final products.

2.3. Integration of AI into additive manufacturing

2.3.1. *Introduction*

The integration of AI into AM opens wide horizons for the optimization and improvement of each stage of the process. While AM continues to gain popularity and extend to new sectors, AI offers innovative solutions for addressing technical challenges and improving operational efficiency.

This section provides an in-depth exploration of the potential of AI in AM, highlighting its advantages and promising applications (Goodfellow 2016; Qiu 2023).

2.3.2. *Feasibility of AI-based solutions*

In AI, assessing the feasibility of AI-based solutions is crucial. Several criteria and characteristics determine if a problem is appropriate for an AI-based solution. First, data availability is an essential factor. AI algorithms often depend on large volumes of data in order to function efficiently, so problems that involve abundant and high-quality data are generally more suited to AI solutions. Furthermore, the nature of the problem itself plays an important role. The problems that involve complex tasks of decision-making, pattern recognition or prediction are often well adapted to AI. Moreover, the possibility to formulate the problem in algorithmic terms and to decompose it into logical stages is also an important aspect of AI feasibility. Finally, the availability of adequate information resources, including the computation power and the required infrastructure, is an essential factor for the efficient resolution of problems using AI. Evaluating these factors helps determine the feasibility of applying AI to a given problem, which is vital for successfully integrating AI into fields such as AM.

2.3.3. *Training data*

In the field of AM, the integration of AI is a promising path for enhancing efficiency, quality and reliability. However, training data are essential for the development of AI models. This section explores in detail the various aspects of data gathering for training AI models in the specific context of AM.

When developing a model for AM, various input data are required for AI model training. This data includes material and geometric properties at different stages of the process. For instance, material data such as thermal and mechanical properties need to be collected from sources such as manufacturer specifications, laboratory tests and specialized online databases. Similarly, geometric data of the parts to be manufactured and process parameters, such as melting temperature and printing speed, are necessary in this case. For example, predicting thermal distortions in a powder bed fusion process involves gathering data on the thermal properties of the powder material used, as well as on the geometric characteristics of the parts being manufactured and the specific process parameters such as laser power and scan speed.

Data collection for training AI models in AM can be complex due to the diversity of data sources and the need for multidisciplinary expertise. The variability in materials, part geometries and manufacturing conditions adds to the complexity of this process. Furthermore, challenges may arise from the limited availability of high-quality data and the need for precise recording to ensure effective AI model training.

2.3.4. *Prediction models and selection of the proper solution*

It is essential to evaluate standard approaches, such as finite element (FE) methods, against emerging approaches such as AI models and hybrid FE–AI models. Traditional FE models, while well established, often require substantial computational resources and time.

In contrast, AI models can predict physical behavior more rapidly, but their effectiveness is heavily dependent on the quality of the training data. Hybrid FE–AI models aim to leverage the strengths of both approaches, though their implementation can be complex. Selecting the proper solution involves considering the specific requirements of the problem, including accuracy, computational efficiency and the availability of data.

In the context of AM, the use of a hybrid FE–AI model may improve the prediction of properties of the manufactured parts. For instance, the FE model can simulate the manufacturing process and predict local effects and properties within the printing area. Meanwhile, the AI model can adjust process parameters in real time based on globally observed results, such as a set of simulation outcomes for the entire part. This approach aims to optimize both computation time and the accuracy of results, thereby minimizing errors and improving the quality of the predictions.

The next part of this chapter delves into the development of AI specifically tailored for AM. We will examine the various stages involved in AI development, including data collection and preparation, model design and training. Additionally, we will address the unique challenges faced during the development of AI for AM and explore future opportunities and perspectives in this field.

2.4. Development of an AI model for additive manufacturing

2.4.1. *Introduction*

This section explores how a prediction model based on AI can be developed for the AM process, with a focus on optimizing process parameters. The goal is to showcase AI's potential in predicting the properties of manufactured parts and making real-time adjustments to process parameters to enhance part quality. Each stage of developing this AI model for AM will be thoroughly examined, including the associated challenges and opportunities for its implementation. By integrating advanced machine learning (ML) techniques with insights into AM technologies, we aim to fully leverage the synergy between these fields to advance AM processes and explore new possibilities in this rapidly evolving domain.

2.4.2. *Machine learning and deep learning*

In the field of AI, machine learning (ML) and deep learning (DL) represent two distinct approaches. ML (see Figure 2.3(a)) involves algorithms that learn from data to develop models without being explicitly programmed for specific tasks. For example, in AM, an ML model might be developed to predict the quality of manufactured parts based on process parameters such as temperature, print speed or material type. In contrast, DL (see Figure 2.3(b)) is a subset of ML that employs deep neural networks to learn data representations through multiple hierarchical layers. For AM, DL techniques could be used to analyze images of manufactured parts to automatically detect printing failures or anomalies, using deep convolutional neural networks (CNN).

The distinction between ML and DL highlights the range of methods available for leveraging data to enhance processes in AM, each with its own strengths and applications.

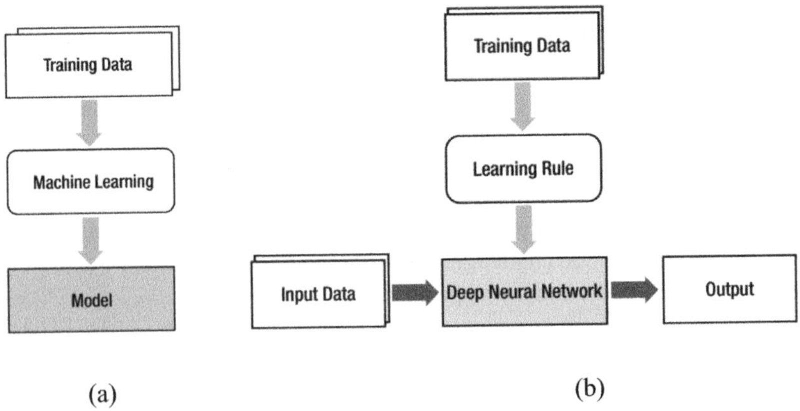

Figure 2.3. *Concepts of machine learning (a) and deep learning (b)*

2.4.3. *Data preparation for training an AI model*

As previously discussed, the quality and relevance of training data are crucial for developing effective AI models. To ensure the robustness and generalization of models, it is essential to rely on high-quality datasets that accurately represent the diversity and complexity of the problems being addressed. Additionally, creating reusable models is a key objective, as it allows for broader applicability and enhances the value of the AI solutions developed. The following stages outline the preparation of training data for our AI model (see Figure 2.4), emphasizing the significance of each step in constructing reliable and high-performance models.

Several stages of data preparation are necessary before training AI models. These include sampling, normalization, standardization, mixing and dividing raw data into training and test sets.

Normalization is an important stage for accelerating the training and improving model generalization. The input variables are generally normalized so that they can be arranged within a specific range, while the outputs can also be normalized according to the model needs. Once the data are prepared, they are randomly mixed to ensure that the model is not biased by the order of samples. Finally, the data are divided into training and test sets, according to a predefined ratio, allowing for the evaluation of model performance on unseen data.

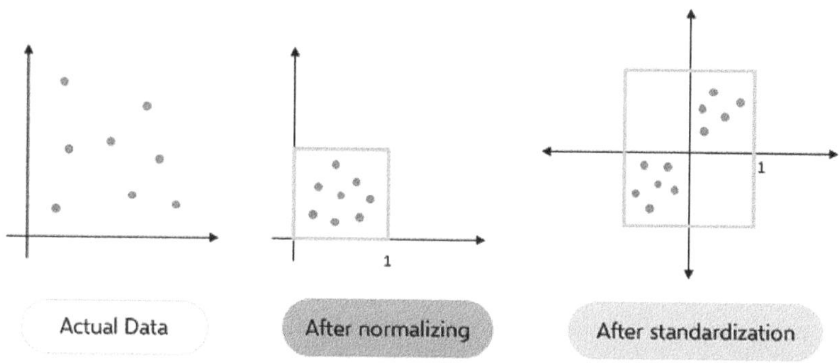

Figure 2.4. *Stages of data preparation for AI model training. For a color version of this figure, see www.iste.co.uk/elhami/applicationsofai.zip*

2.4.4. *AI model training*

Once the data are prepared, the AI model is trained on a training set using a specific method according to the chosen model type. For supervised learning models, such as artificial neural networks (ANN) (Figure 2.5) or support vector machines (SVM), input data are provided to the model, which adjusts its parameters for minimizing a loss function, often the mean-squared errors (MSE). Training is generally conducted over several periods, each of which consists of a full pass through the training set. During the training, the model performance is regularly assessed on a validation set to monitor for overfitting and to ensure that the model generalizes well to unseen data.

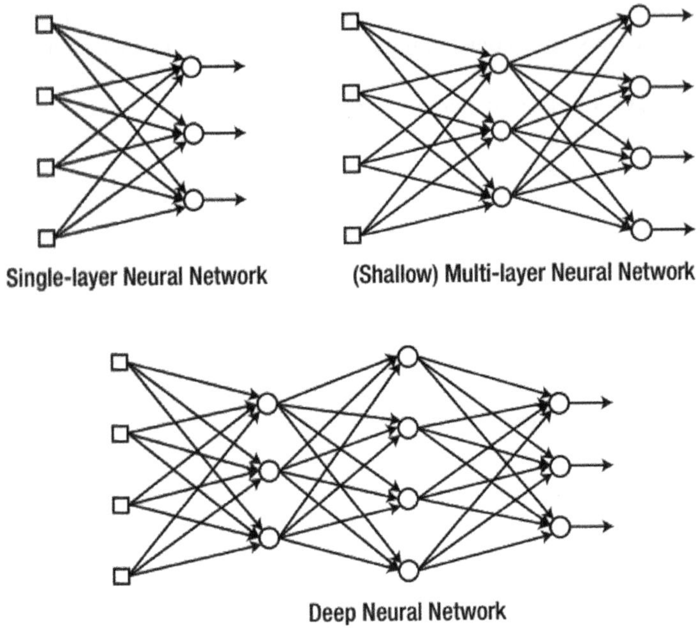

Figure 2.5. *Types of artificial neural networks*

The previously introduced delta rule is not efficient for training a multi-layer neural network. This is because the error, the essential element for applying the delta rule for training, is not defined in the hidden layers. The output node error is defined as the difference between the correct output and the neural network output. However, training data do not provide correct outputs for the nodes of the hidden layers; therefore, the error cannot be calculated using the same approach for the output nodes.

So, what is the solution? The essential issue lies in defining the error for the hidden nodes. This challenge is precisely addressed by the backpropagation algorithm (Figure 2.6), which is the learning rule tailored for multi-layer neural networks.

The introduction of the backpropagation algorithm finally solved the problem of training the multi-layer neural network. The significance of the backpropagation algorithm was that it provided a systematic method for determining the errors on hidden nodes. Once the errors on the hidden layer are determined, the delta rule is applied to adjust weights.

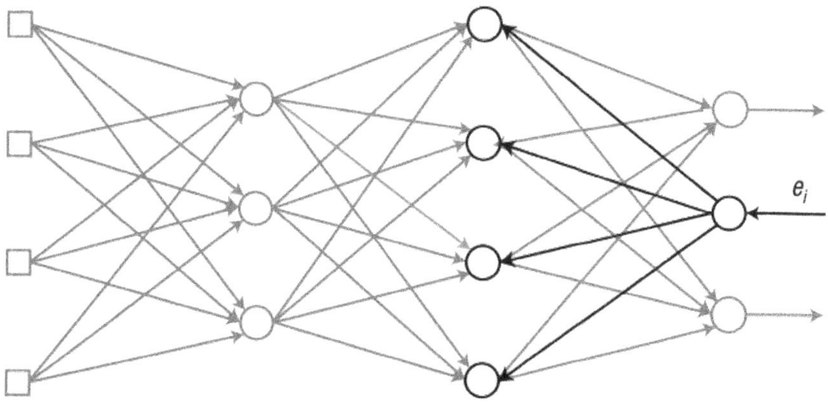

Figure 2.6. *Illustration of backpropagation*

2.4.5. *Performance of the AI model prediction*

Once the model is trained, its performance is assessed on a different test set to evaluate its capacity to generalize on new data. Metrics such as mean-squared error (MSE), the coefficient of determination (R2) and other indicators are often used to evaluate the precision of the model on test data. A high performance of the model is indicated by a low error and a strong ability to accurately predict results on unseen data.

2.4.6. *Application to the optimization of manufacturing parameters*

2.4.6.1. *Training data*

In order to develop an AI model aiming to optimize the AM parameters, a variety of data can be used. This data may include process parameters such as temperatures, extrusion rate, the scan speed, the fill rate and other parameters specific to the AM technology.

Additionally, it is essential to include specific characteristics of the materials employed, such as the mechanical and thermal properties, density and surface roughness. Moreover, data on the performance of the manufactured parts, used for quality control depending on the type of part produced, such as strength for mechanical components, are also required for the efficient training of the AI model.

Several methods can be employed to obtain this data. In a practical context, sensors integrated into 3D printing machines can collect data during the manufacturing process. These sensors can measure parameters such as nozzle temperature, printing zone temperature, print speed and the dimensions of the deposited layers.

Additionally, monitoring devices can be used to gather data on the performance of the manufactured parts, such as results from mechanical strength tests or surface analyses.

By combining this data with historical datasets from previous manufacturing processes, an AI model can be trained to predict the optimal AM parameters based on the specific characteristics of each part and the expected performance outcomes.

2.4.6.2. *AI model for the optimization of AM parameters*

The model presented in Figure 2.7 illustrates the optimization process of AM parameters:

– Input: this part of the model represents the input data used for training and prediction. In the case of optimization of the AM parameters, the input data include the specific process parameters that we aim to optimize, such as the printing temperature, the extrusion rate, the fill rate, etc., as well as the characteristics of the materials employed (El Hami 1966; Kharmanda 2014).

– AI model: this section of the model represents the AI algorithm used to process input data and generate predictions. This can be a supervised or unsupervised ML model, or a DL model such as a convolutional neural network (CNN) or a recurrent neural network (RNN).

– Output: this is the part of the model that represents the results, or the predictions generated by the AI model. In the context of optimization of the parameters of AM, the output can be the AM parameters optimized to obtain the highest performances of manufactured parts.

– Error: this last component of the model represents the difference between the results predicted by the model and the monitoring data gathered during the manufacturing process or after the manufacturing of the parts. These data are used to assess the actual performance of the manufactured parts and can include measures such as the mechanical strength, the quality of the surface, etc. The error is an essential indicator that can be used to adjust and continuously improve the AI model to obtain more accurate predictions and optimal manufacturing results.

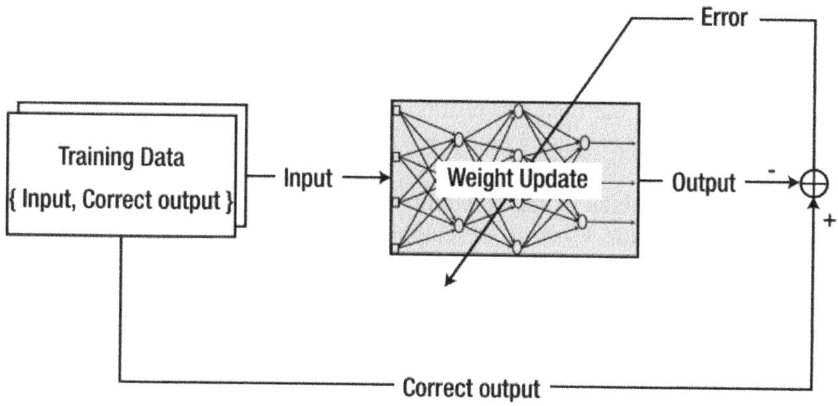

Figure 2.7. *Concept of supervised learning*

To summarize, the development of an AI model for the optimization of AM parameters represents significant potential in the field of AM. Using data from various processes and the performance control data of various parts, it is possible to develop a model that can be used to identify the optimal parameters that ensure high-quality manufacturing results. By integrating ML and DL techniques, this model can adapt and continuously improve to meet the changing demands of manufacturing activities. Thanks to this approach, the companies can reduce costs, enhance efficiency and ensure the quality of manufactured parts, thus opening a path to new possibilities in various industrial sectors.

2.5. Conclusion

The current study highlights the various ways in which AI can revolutionize AM. The summary of the potential advantages of AI in this field and the call to action for future research help us contemplate a future in which AI radically transforms the manners in which we design, manufacture and use products thanks to AM.

AI can optimize certain stages of AM through an advanced use of data analysis and ML. It can analyze large amounts of data on the existing design, the performances of materials and the functional requirements to propose more efficient and innovative design concepts. Similarly, during the manufacturing phase, it can adjust the manufacturing parameters such as temperature, print speed and fill density depending on the characteristics of each part and the expected performances, relying on models built based on data gathered in real time or on data from previous manufacturing processes.

AI can progressively improve the quality and efficiency of AM while allowing dynamic adaptation to variations and challenges encountered before, during and after the manufacturing phase. By identifying models and relations between the process variables and the obtained results, it can guide design and manufacturing decisions to obtain higher-quality products by reducing development time and costs.

To take full advantage of the potential of AI in AM, it is essential to continue research in several fields. First, the development of new learning techniques adapted to the specific needs of AM is crucial, particularly regarding the management of massive data and the consideration of manufacturing constraints. Second, it should be studied how to efficiently integrate AI into existing manufacturing processes, considering practical challenges and industrial requirements. Finally, it is important to explore new fields of application for AI in AM, such as the design of complex structures and product customization.

In conclusion, we can consider a future in which AI radically transforms AM, enabling the creation of more innovative, sustainable products that are better adapted to users' individual needs. Thanks to the continuous progress in research and development, AI is expected to revolutionize the way we design and manufacture objects, thus paving the way for a new age of digital manufacturing.

2.6. References

Abo Al-Kheer, A., Eid, M., Aoues, Y., El-Hami, A., Kharmanda, M.G., Mouazen, A.M. (2011). Theoretical analysis of the spatial variability in tillage forces for fatigue analysis of tillage machines. *Journal of Terramechanics*, 48(4), 285–295.

Bahrin, M.A.K., Othman, M.F., Azli, N.H.N., Talib, M.F. (2016). Industry 4.0: A review on industrial automation and robotic. *Jurnal Teknologi*, 78, 6–13. doi: 10.11113/jt.v78.9285.

Ben Abdessalem, A. and El Hami, A. (2015). A probabilistic approach for optimising hydroformed structures using local surrogate models to control failures. *International Journal of Mechanical Sciences*, 96, 143–162.

Ben Abdessalem, A., Pagnacco, E., El Hami, A. (2013). Increasing the stability of T-shape tube hydroforming process under stochastic framework. *International Journal of Advanced Manufacturing Technology*, 69(5–8), 1343–1357.

Bouyekhf, R., El Moudni, A., El Hami, A., Zerhouni, N., Ferney, M. (1996). Reduced order modelling of two-time-scale discrete non-linear systems. *Journal of the Franklin Institute*, 333(4), 499–512.

Bouyekhf, R., El Hami, A., El Moudni, A. (2001). Optimal control of a particular class of singularly perturbed nonlinear discrete-time systems. *IEEE Transactions on Automatic Control*, 46(7), 1097–1101.

Chua, C.K., Leong, K.F., Lim, C.S. (2010). *Rapid Prototyping: Principles and Applications*, 3rd edition. World Scientific Publishing Company, Singapore. doi: 10.1142/6665.

El Hami, A. and Radi, B. (1996). Some decomposition methods in the analysis of repetitive structures. *Computers and Structures*, 58(5), 973–980.

El Hami, A., Lallement, G., Minotti, P., Cogan, S. (1993). Methods that combine finite group theory with component mode synthesis in the analysis of repetitive structures. *Computers and Structures*, 48(6), 975–982.

Gibson, I., Rosen, D.W., Stucker, B. (2010). *Additive Manufacturing Technologies: Rapid Prototyping to Direct Digital Manufacturing*. Springer, Heidelberg. doi: 10.1007/978-1-4419-1120-9.

Goodfellow, I., Bengio, Y., Courville, A. (2016). *Deep Learning*. MIT Press, Cambridge. doi: 10.7551/mitpress/9780262035613.001.0001.

Huang, S.H. and Liu, P. (2012). Additive manufacturing and its societal impact: A literature review. *The International Journal of Advanced Manufacturing Technology*, 67(5–8). doi: 10.1007/s00170-012-4558-5.

Kharmanda, G., Ibrahim, M.-H., Abo Al-kheer, A., Guerin, F., El-Hami, A. (2014). Reliability-based design optimization of shank chisel plough using optimum safety factor strategy. *Computers and Electronics in Agriculture*, 109, 162–171.

Kim, P. (2017). *MATLAB Deep Learning: With Machine Learning, Neural Networks and Artificial Intelligence*. Apress, New York. doi: 10.1007/978-1-4842-2845-6.

LeCun, Y., Bengio, Y., Hinton, G. (2015). Deep learning. *Nature*, 521, 436–444. doi: 10.1038/nature14539.

Mabrouk, I.B. and El Hami, A. (2019). Effect of number of blades on the dynamic behavior of a Darrieus turbine geared transmission system. *Mechanical Systems and Signal Processing*, 121, 562–578.

Makhloufi, A., Aoues, Y., El Hami, A. (2016). Reliability based design optimization of wire bonding in power microelectronic devices. *Microsystem Technologies*, 22(12), 2737–2748.

Moro, T., El Hami, A., El Moudni, A. (2002). Reliability analysis of a mechanical contact between deformable solids. *Probabilistic Engineering Mechanics*, 17(3), 227–232.

Ngo, T.D., Kashani, A., Imbalzano, G., Nguyen, K.T.Q., Hui, D. (2018). Additive manufacturing (3D printing): A review of materials, methods, applications and challenges. *Composites Part B: Engineering*, 143, 172–196. doi: 10.1016/j.compositesb.2018.02.012.

Qiu, J. and Jiang, L. (2023). Development of modular and reusable AI models for fast predicting fire behaviour of steel columns in structural systems. *Engineering Structures*, 297, 116994. doi: 10.1016/j.engstruct.2023.116994.

3

Optimization of BGA Components under Real Service Conditions

3.1. Introduction

The reliability of electronic devices is largely dependent on the solder joints connecting the electronic components with the printed circuit boards. It is particularly crucial for BGAs (Ball Grid Arrays) that integrate multiple contacts within a limited zone. The reliable design of solder joints is essential for reaching optimal performances and longevity. If electro-thermomechanical operating conditions are ignored during the design process, significant failures of various types may occur. The design of BGA solder joints may be a difficult task, as their reliability is strongly influenced by many factors, such as the geometry and properties of materials. For a BGA component to reach optimal performance, there needs to be thorough consideration of the various critical parameters, including the solder ball diameter, the solder volume and the height of the case. The size of the BGA component, the required amount of solder and its thermal and mechanical properties will be strongly affected by the solder ball diameter.

According to previous research projects conducted by Sinda et al. (2023, 2024), the modification of the radius and height of BGA components may have a significant impact on their reliability and fatigue life. The appropriate selection of the solder volume is a fundamental prerequisite for optimal performances, as an insufficient quantity of solder may lead to weak connections, while an excessive quantity may lead to short-circuits and other problems. Previous research on the

Chapter written by Sinda GHENAM, Abdelkhalak EL HAMI, Wajih GAFSI, Ali AKROUT and Mohamed HADDAR.

reliability of ball grids mainly focused on various aspects such as modeling, optimization, defect analysis and service life assessment. The work conducted by Müller and Hauck (2008) examines the reliability of microelectronic soldering based on the combination of experiments and data analyses. They also used a thermomechanical database to examine the behavior of soldering materials under various thermal cycle conditions.

The main objective of this chapter is the optimization of the BGA component. From this perspective, we are reviewing various optimization techniques presented in the literature. In this respect, Lau et al. (2013) adopted the Taguchi method to optimize the reflow soldering process and have shown that design optimization can reduce thermomechanical stresses on the soldering joints, thus improving their reliability. In this context, it is important to note that the Taguchi method only requires nine experiments to find the optimal conditions, while RSM requires 20, which saves approximately 55% of time and effort, according to Uday Kiran et al. (2021). In this context, Canumalla et al. (2008) and Zhang et al. (2013) used the response surface methodology (RSM) to solve reliability and robust design problems in advanced packaging. Zhang et al. (2013) used RSM in their study of temperature distribution of a multilayer module. They conducted a thermal optimization by adjusting the parameters depending on the regression equation to minimize the maximal junction temperature of the chip. Finally, Hadim and Suwa (2004) and Fraternali et al. (2009) applied a genetic algorithm to optimize several design parameters. These studies highlighted the active research field of optimization methods, and in particular, the optimization of the electro-thermomechanical reliability of BGA components, with various methods used for minimizing thermomechanical stresses and improving reliability. However, each technique has its own limitations that need to be carefully considered with respect to its application to our problem specifications. To achieve a properly balanced optimal design, several criteria such as reliability, sustainability, feasibility and profitability must be taken into account. Hence, the basic objective of the selection of a well-adapted optimization technique is to reach a better balance between cost reduction and performance enhancement. It is in this context that this work was conducted to evaluate the use of the reliability-based design optimization (RBDO) method for optimization applications. By coupling the finite element model developed by the Ansys Parametric Design Language (APDL) software and the RBDO model coded with MATLAB software, we are able to specify the optimal variables that can improve the efficiency of the structure. Our approach also takes into account the economic feasibility of the BGA structure alongside its technical aspects. By incorporating these criteria, our methodology provides designers with a means to achieve the optimal design of the BGA component, which enables the enhancement of reliability and efficiency, while keeping costs under control.

3.2. Presentation of the optimization method

In order to improve the performance of the electronic devices while reducing their size, various geometry optimization approaches can be considered. Commonly used strategies include the optimization of materials, shape, layout, location and size. Shape optimization involves modifications to improve performance while location optimization aims for more precise readings by modifying the position of components. Layout optimization involves the minimization of electromagnetic interferences through an appropriate arrangement of components. Size optimization reduces dimensions while maintaining performances, while the optimization of materials uses more reliable materials, such as those with higher thermal conductivity, to optimize heat dissipation. These methods can be used as such or in combination, in order to optimize the performance and the dimensions of electronic components.

Our research focuses on the optimization of joint geometry in order to improve their reliability, taking into account the actual operating conditions. The main objective is to ensure a high reliability level throughout the service life of the device, while meeting the criteria defined by the designer. From this perspective, we used RBDO algorithms to obtain optimal design variables, as well as a targeted reliability index. These algorithms aim to minimize an objective function, represented here by the soldering joint volume, in order to reduce the amount of material used. First, we determined the initial geometry of the BGA component, in particular the initial radius and height, based on a technical specification sheet from Infineon Technologies. Simulations were then conducted using Ansys APDL software to obtain the cumulative plastic deformation of soldering joints. This deformation can be used to identify the areas of concentration of stresses and deformations, as well as the potential beginning of the failure, which is essential for our study. The design variables, namely the radius and height, as well as the cumulative plastic deformation, are considered as the main input data of the optimization algorithm, which operates through an Ansys and MATLAB interoperability loop. These variables must meet certain conditions: they must range between the higher and lower limits defined in the literature, maintain the cumulative plastic strain below the maximal accepted limit, minimize the volume of the soldering joint and reach a target reliability index set at $\beta t = 3$. If the obtained results meet these criteria, then we have achieved our objective. Otherwise, we adjust the geometric parameters until all specified conditions are met.

The selection of the RBDO methodology represents a preferred option for reaching optimal and reliable solutions in the engineering field. This approach stands out because of its capacity to consider the nonlinearity of the studied systems, which contributes to a more accurate comprehension of the complex behaviors of the components. Moreover, unlike other methods, RBDO acknowledges the dependence between various design parameters, which contributes to a more realistic modeling of the systems. Finally, RBDO often proves to be a less expensive option compared to other approaches, thus offering an economical solution to reach optimal and reliable designs.

3.3. Presentation of the system

In this chapter, we decided to focus on the study of the ball grid array (BGA) for our research. BGA is widely used in modern electronics due to its multiple advantages, such as a high connection density, enhanced thermal dissipation and increased resistance to vibrations and impacts. Concretely, it is composed of a substrate on which there are solder balls, forming a network of connections through which electric signals are transferred and heat is dissipated.

In the context of our work, our assembly has a complex structure, composed of several layers of different materials. The silicon chip, the copper layers representing the metallic junction and the FR4 layer of the printed circuit act jointly to ensure the overall proper operation. To model this assembly, we used ANSYS APDL. In order to optimize the simulation time while ensuring accurate results, we decided to model only one-quarter of the complete geometry. The mesh was selectively refined to concentrate the highly refined elements in the solder joints, where failures are more likely to occur.

The dimensions and technical specifications of our assembly were extracted from an industrial technical data sheet. The size of the chip was 3 × 5 × 0.55 mm, while the copper layers had a dimension of 3 × 5 × 0.36 mm. Similarly, the spacing height of the solder joints was 0.2 mm, and the FR4 layer of the printed circuit measured 3 × 5 × 0.2 mm.

As for the boundary conditions, we applied two symmetries on the opposite surfaces of silicon, copper and of the printed circuit for displacements. Moreover, for thermal boundary conditions, we have ignored the radiation phenomenon and have assumed a thermal convection coefficient "h" ranging between 10 and 25 W m-2 K-1, given that we only consider natural convection. The global description is presented in Figure 3.1.

Given the importance of the impact of various materials in the electro-thermomechanical simulation of the system, a detailed analysis of the properties of materials of each part of the assembly is essential. These properties are extracted from various sources in the literature, particularly the thesis of Halouani et al. (2020), the study of Van Nhat (2016), and also the works of Sinda et al. (2023).

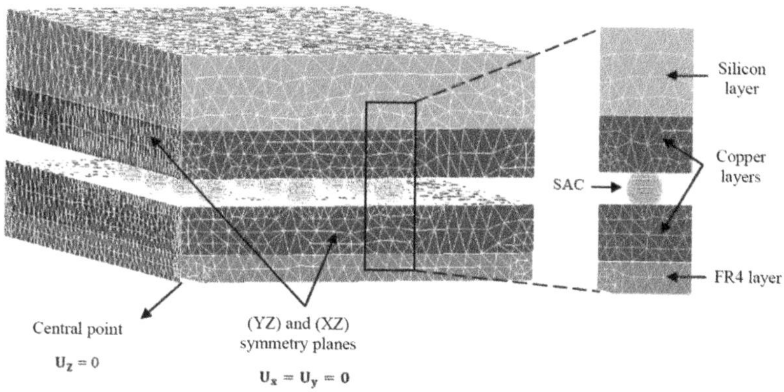

Figure 3.1. *Global model of a quarter of the BGA assembly with boundary conditions. For a color version of this figure, see www.iste.co.uk/elhami/applicationsofai.zip*

3.4. Methodology

3.4.1. *Deterministic design optimization (DDO)*

The DDO method aims to minimize the brazing joint volume while ensuring its reliability. This means that the accumulated plastic strain, responsible for the failure, must remain below a maximal limit. The design variables of this optimization are the radius "r" and the height "h". Uncertainties are commonly considered by applying a safety factor S_f to the plastic strain ε_p of the soldered joint, which is often determined according to the engineer's experience. In our case study, this safety factor was set at 1.5, a value approved by the commercial industries and supplied by the "engineering toolkit". Consequently, the maximal plastic strain in the joint should not exceed 0.167.

Our formulation of the optimization problem is as follows:

$$\begin{aligned} &\text{Min } V(r,h) = \frac{1}{3} * \pi * r^2 * h \\ &\text{S.C } \varepsilon_p < \frac{s_{p_{adm}}}{S_f} \\ &r_L < r < r_H \\ &h_L < h < h_H \end{aligned} \qquad [3.1]$$

3.4.2. *Reliability-based design optimization (RBDO)*

This type of optimization offers the possibility to have a compromise between reaching the expected reliability level and meeting the design constraints, which contributes to obtaining more robust and efficient design solutions. There are several methods in this optimization category, as presented in Figure 3.2. These methods will be presented and applied in the rest of this chapter.

These methods aim to efficiently balance the safety and profitability in the design of our BGA components.

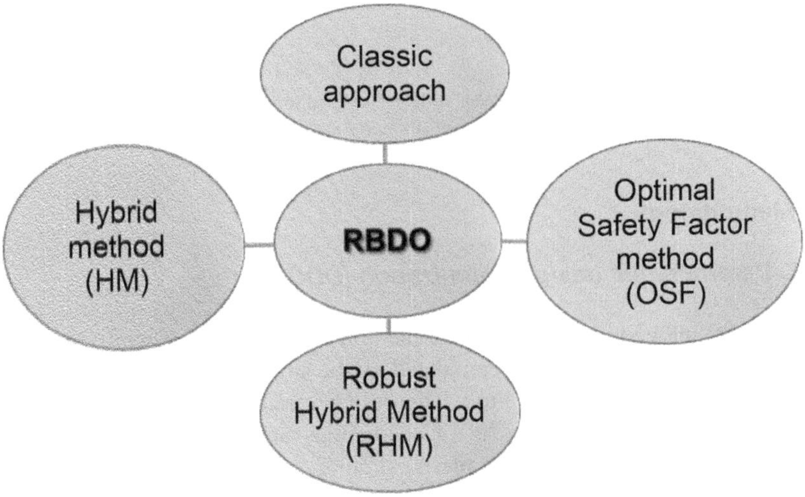

Figure 3.2. *Various RBDO methods. For a color version of this figure, see www.iste.co.uk/elhami/applicationsofai.zip*

3.4.2.1. *OSF method*

The approach of this method is described in three main steps as shown in Figure 3.3.

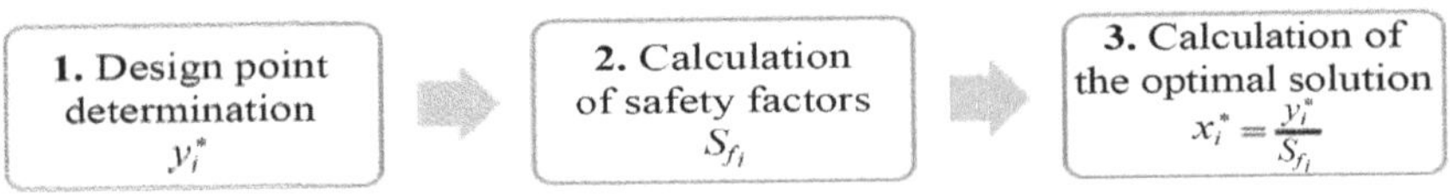

Figure 3.3. *Approach of the OSF method*

To reach the optimal solution, it is first necessary to determine the design point. The latter is identified by solving the optimization problem described in Figure 3.4:

$$\begin{aligned} &\min_{y} \; V(y) \\ &\text{S.C} \;\; \varepsilon_{p_{adm}} \leq \varepsilon_{p_{cri}} \\ &y_L < y < y_U \end{aligned} \qquad [3.2]$$

3.4.2.2. *Hybrid method*

This approach has at its core a search for a new formulation of the objective function, aiming to optimize the function F (x, y) while meeting the deterministic constraints g(x), the reliability constraint and the performance function G (x, y). The variable d_β (x, y) represents the distance between the optimal solution and the design point. In a hybrid design space (HDS), the objective function is minimized by taking into account both the deterministic parameters, x, and the random parameters, y. An illustration of the HDS for two random variables with a normal distribution is presented in Figure 3.4. This figure shows that the reliability levels d_β can be represented by ellipses, while the optimal solution, P_x^*, and the most probable point, P_y^*, can be identified at the intersection of the curve G (x, y) = 0 and $d_\beta = \beta_t$.

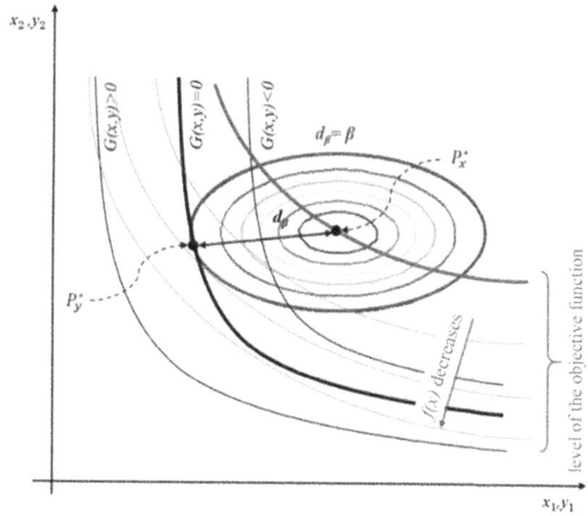

Figure 3.4. *Hybrid design space (HDS). For a color version of this figure, see www.iste.co.uk/elhami/applicationsofai.zip*

Its mathematical formulation is similar to the following:

$$\begin{aligned}
&\min_{x,y} \ V(r,h)*d_\beta(x,y) \\
&\text{S.C} \ \ \varepsilon_{P_{adm}} \leq \varepsilon_{P_{cri}} \\
&d_\beta(x,y) \geq \beta_t \\
&x_L < x < x_U \\
&y_L < y < y_U
\end{aligned} \qquad [3.3]$$

The classical hybrid method (HM) has a major drawback because of the significant computational time it involves and its tendency to converge to a solution, P_x^*, located in the dangerous area (G(x) > 0), even when the distance $d_\beta = (x, y)$ ensures an acceptable reliability level, as illustrated in the figure.

3.4.2.3. *Robust hybrid method (RHM)*

To address this situation, Yaich et al. (2020) proposed a solution known as RHM. This approach integrates a constraint expected to ensure that the optimization algorithm discovers the optimal solution in the safe zone.

Consequently, in the context of RHM, the value of the objective function at the optimal solution, P_x^*, must be above that of the design point P_y^*.

Therefore, RHM can be defined as follows:

Minimizing *x*,*y* the function $F(x,y) = f(x)*d_\beta(x,y)$

Under the constraints: $G(x,y) \leq 0$

gk (x) ≤ 0, k = 1,..,K

$d_\beta(x,y) \geq \beta_t$

$$f(x) \geq f(y) \qquad [3.4]$$

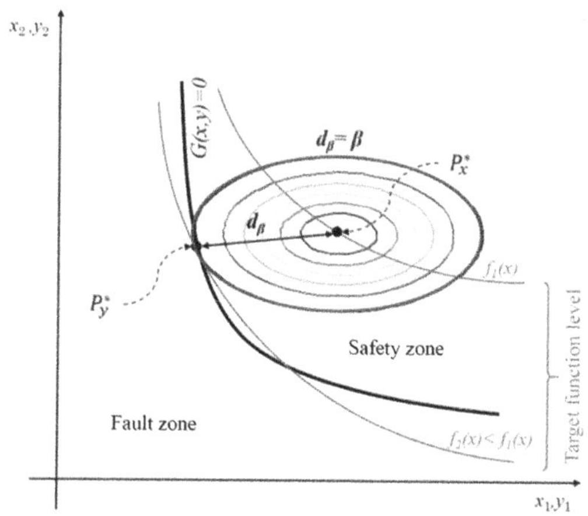

Figure 3.5. *Efficiency of the RHM method. For a color version of this figure, see www.iste.co.uk/elhami/applicationsofai.zip*

Figure 3.5 illustrates an example of RHM in HDS, where two objective functions, f1(x) and f2(x), meet with the optimal solution P_x^*, and the design point P_y^*. The optimal point, P_x^*, is located in the safe region, where $f(x*)$ is above $f(y*)$, and the reliability constraint, $d_\beta = \beta_t$, is also met. In the figure, the diagram presents the entire method RHM, which proved an enhancement in robustness and a reduction in the computational time compared to the standard HM method. The results of the numerical simulation of the BGA component subjected to electro-thermomechanical loads in the ANSYS APDL software will be presented in the next section and then combined with MATLAB for analysis and discussion.

3.5. Results

3.5.1. *DDO*

Based on the analysis of the results of the DDO, we are able to predict the optimal configuration, denoted by X_{opt}, for which the minimal value of the volume, V_{opt}, can be reached. Moreover, it can be noted that the various constraints were met. The design variables are well within fixed intervals: the radius, r, ranges between [0.15e-3; 0.375e-3] and the height, h, ranges between [0.2e-3; 0.8e-3], similarly to the optimization constraint, $\varepsilon_{p\ max} < 0.167$. The figure shows a graphical representation of the efficiency of the DDO approach in terms of optimization of the geometry. This figure compares the curves of the calculated initial and optimal constraints, highlighting a clear improvement in the accumulated plastic strain. This is reduced by $\Delta\varepsilon acc$ = 15.4 %, according to equation [3.5]:

$$\Delta\varepsilon acc = \frac{\varepsilon_{acc\ max} - \varepsilon_{acc\ initial}}{\varepsilon_{acc\ max}} = \frac{0.162 - 0.137}{0.162} = 0.154 = 15.4\% \quad [3.5]$$

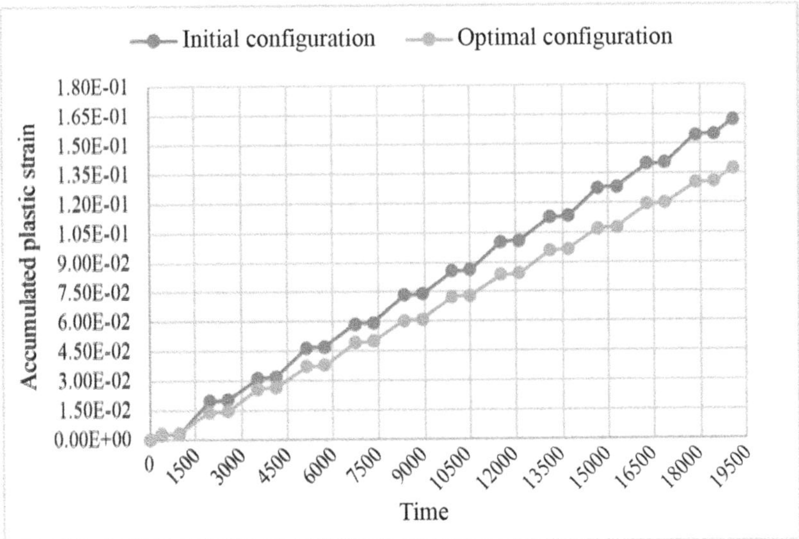

Figure 3.6. *Comparison of the initial and optimal configurations. For a color version of this figure, see www.iste.co.uk/elhami/applicationsofai.zip*

Our results show a significant improvement in the accumulated plastic deformation, which confirms the efficiency of our design approach. Indeed, the solution obtained meets both the plastic deformation constraint and the intervals of design variables, thus validating the quality of our optimal design. However, despite this success, it is crucial to note that the reliability index does not meet the intended and expected levels, fluctuating between higher and lower levels of the target reliability index. This situation may potentially lead to an unsafe solution, thus increasing the risk of component failure. This absence of reliability constraint is the main limitation of the deterministic design method used, even by taking into account safety factors deduced from engineering experiences. Consequently, the adoption of RBDO becomes of great interest when reaching reliability levels closer to expectations. By using this approach, we can obtain more robust and efficient design solutions, thus meeting stricter demands in terms of reliability of electronic components.

3.5.2. RBDO

3.5.2.1. *OSF*

When the optimal values x_i^* are outside the required range, this can compromise the stability of the optimal safety factor (OSF) method in some situations. To address this problem and have an efficient resolution, it is imperative to consider using the HM. Using the HM method enables us to overcome the challenges due to the optimal values outside the required range, thus ensuring the stability and efficiency of the optimization process.

3.5.2.2. *Hybrid method: HM*

The objective function of the HM problem is expressed in terms of the product of volume and distance in the HDS:

$$\begin{aligned}
&\min_{x,y} \; F(x,y) = V(r,h) * d_\beta(x,y) \\
&\text{S.C } \; \varepsilon_{p_{adm}} \leq \varepsilon_{p_{cri}} \\
&d_p(x,y) \geq \beta_t \\
&x_L < x < x_U \\
&y_L < y < y_U
\end{aligned} \qquad [3.6]$$

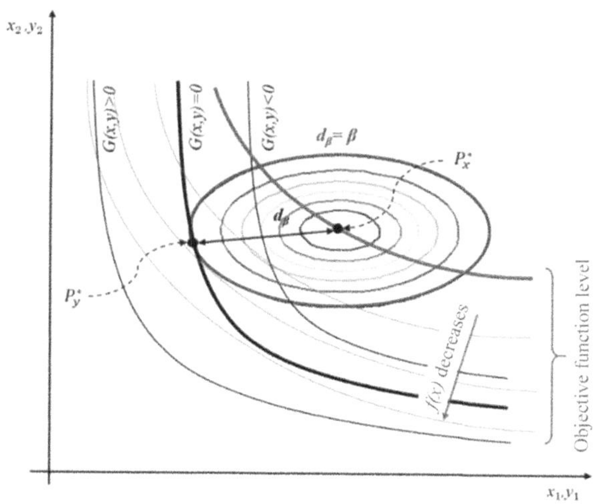

Figure 3.7. *Hybrid design space. For a color version of this figure, see www.iste.co.uk/elhami/applicationsofai.zip*

Following an in-depth analysis, we concluded that the maximal plastic deformation, represented by $\varepsilon_{p\ max}$, is located in an unsafe region, exceeding the critical value of 0.25. This raises concerns related to the reliability of our design. However, by examining the distance between the design point and the optimal point, represented by β, an acceptable reliability level can be noted. To address this problem and ensure that the optimal point is in a safe zone, we have opted for a robust hybrid method (RHM). This method efficiently combines robust techniques with standard optimization methods, thus ensuring the reliability and stability of our optimal solution.

3.5.2.3. *Robust hybrid method: RHM*

At this stage, the optimal solution can be safely found in the verified reliability zone, with design variables rigorously respecting predefined limits, particularly for the optimization constraint $\varepsilon_{p\ max} < 0.25$

$$\begin{aligned}
&\min_{x,y}\ F(x,y) = V(r,h) * d_\beta(x,y) \\
&\text{S.C}\ \varepsilon_{p_{adm}} \le \varepsilon_{p_{cri}} \\
&d_\beta(x,y) \ge \beta_t \\
&V(x) \ge V(y) \\
&x_L < x < x_U \\
&y_L < y < y_U
\end{aligned} \qquad [3.7]$$

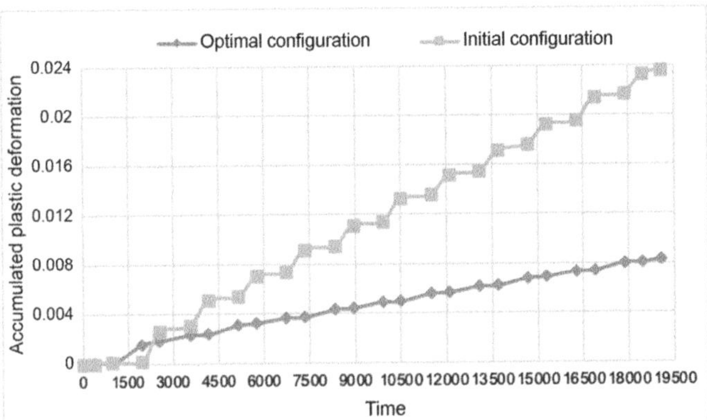

Figure 3.8. *Comparison of the initial and optimal configurations. For a color version of this figure, see www.iste.co.uk/elhami/applicationsofai.zip*

A significant improvement of the accumulated plastic deformation can also be noted, with a 33% reduction. This substantial reduction confirms the relevance and optimality of the obtained solution. This success is convincingly highlighted by the illustration of the efficiency of the design optimization approach through the use of the RHM.

Parameters

Radius (r)/height (h)/volume (V)/reliability index ( )

Methods	Parameters	Initial point x_0	Optimal point x_i^*	Results
DDO	R	0.15	0.22	* Optimal solution. * Fluctuating reliability index that leads to an unsafe solution.
	h	0.2	0.38	
	V	0.47 e-11	1.92 e-11	
	$\varepsilon_{p\ max}$	0.0235	0.0095	
	β		4.9	
OSF	r	0.15	0.1457	x_i^* is outside the required range → The OSF method may be unstable in some circumstances.
	h	0.2	0.18	
	V	0.47 e-11	0.4 e-11	
	$\varepsilon_{p\ max}$	0.0235	0.0266	
	β		3	
HM	r	0.15	0.3	* $\varepsilon_{p\ max}$ is in the unsafe region with a maximal value of plastic strain: $\varepsilon_{p\ max} > 0.25$. * β indicates an acceptable level of reliability.
	h	0.2	0.2	
	V	0.47 e-11	1.88 e-11	
	$\varepsilon_{p\ max}$	0.0235	0.32	
	β	3.2	3	

<table>
<tr><td rowspan="5">RHM</td><td>r</td><td>0.15</td><td>0.26</td><td rowspan="5">The optimal solution is in the safe zone → the reliability level is duly checked. The design variables are within the set range, and so is the optimization constraint $\varepsilon_{p\ max}$.</td></tr>
<tr><td>h</td><td>0.2</td><td>0.39</td></tr>
<tr><td>V</td><td>0.47 e-11</td><td>2.76 e-11</td></tr>
<tr><td>$\varepsilon_{p\ max}$</td><td>0.0235</td><td>0.0083</td></tr>
<tr><td>β</td><td>3.2</td><td>3</td></tr>
</table>

Table 3.1. *Optimization results*

3.6. Conclusion

In the industrial and research fields, the crucial importance of reliable and optimally designed BGA components has continuously increased. To address this challenge, we elaborated an innovative approach aimed at optimizing the structure of BGAs. By adopting the RBDO method, we evaluate the geometric design variables of BGA components and determine the optimal parameters supporting structural efficiency. Our approach takes into account both the technical and economic feasibility, thus ensuring an optimal design of BGA solder joints. This study provides detailed information on the results of our research, an in-depth analysis of collected data, while highlighting the advantages of our approach and its potential for practical applications.

To increase the reliability of BGA electronic components, we developed an approach based on a numerical model using finite elements supported by ANSYS software, associated with an optimization algorithm implemented via MATLAB. Our main objective was to reduce the volume of soldered joints while taking into account the plastic deformation constraints and the design variables, in order to obtain optimal BGA soldered joints meeting the reliability demands.

After noting the reliability and safety problems with other reliability-based design methods (RBDO) such as HM and OSF, we implemented a new method known as RHM. This approach stands out because of its remarkable computational efficiency reflected in a substantial improvement of computational time, while ensuring reliability and the accumulated plastic deformation constraint. Our RBDO approach proved to be efficient in solving the encountered optimization problems, which indicates that it is particularly fit for the electronic industries, with promising applications in this field.

Despite its advantages, this work requires time to solve the ANSYS–MATLAB pairing problems. To improve this, our work can be extended by replacing the ANSYS model with a meta-model, not only to improve the pairing, but also to reduce the computation time. In conclusion, this synthesis may serve as a pioneering work for fruitful research and a guideline for future promising research.

3.7. References

Abo Al-Kheer, A., Eid, M., Aoues, Y., El-Hami, A., Kharmanda, M.G., Mouazen, A.M. (2011). Theoretical analysis of the spatial variability in tillage forces for fatigue analysis of tillage machines. *Journal of Terramechanics*, 48(4), 285–295.

Ben Abdessalem, A. and El Hami, A. (2015). A probabilistic approach for optimising hydroformed structures using local surrogate models to control failures. *International Journal of Mechanical Sciences*, 96–97, 143–162.

Ben Abdessalem, A., Pagnacco, E., El-Hami, A. (2013). Increasing the stability of T-shape tube hydroforming process under stochastic framework. *International Journal of Advanced Manufacturing Technology*, 69(5–8), 1343–1357.

Bouyekhf, R., El Moudni, A., El Hami, A., Zerhouni, N., Ferney, M. (1996). Reduced order modelling of two-time-scale discrete non-linear systems. *Journal of the Franklin Institute*, 333(4), 499–512.

Bouyekhf, R., El Hami, A., El Moudni, A. (2001). Optimal control of a particular class of singularly perturbed nonlinear discrete-time systems. *IEEE Transactions on Automatic Control*, 46(7), 1097–1101.

Canumalla, S. (2008). Electro-thermomechanical modeling of a BGA assembly subjected to a damaging displacement and to random vibrations. *Proceedings of the Third International Conference on Advanced Materials, Mechanics, and Manufacturing (A3M'2021)*. IEEE, Lake Buena Vista.

Dammak, K., Yaich, A., El Hami, A., Walha, L., Haddar, M. (2020). An efficient optimization based on the robust hybrid method for the coupled acoustic–structural system. *Mechanics of Advanced Materials and Structures*, 27(21), 1816–1826. doi: 10.1080/15376494.2018.1525629.

El Hami, A. and Radi, B. (1996). Some decomposition methods in the analysis of repetitive structures. *Computers and Structures*, 58(5), 973–980.

El Hami, A., Lallement, G., Minotti, P., Cogan, S. (1993). Methods that combine finite group theory with component mode synthesis in the analysis of repetitive structures. *Computers and Structures*, 48(6), 975–982.

Fraternali, F., Porter, M.A., Daraio, C. (2009). Optimal design of composite granular protectors. *Mechanics of Advanced Materials and Structures*, 17(1), 1–19. doi: 10.1080/15376490802710779.

Ghenam, S. (2024). Fiabilité des systèmes électroniques multiphysiques en environnement critique : application à l'optimisation des composants BGA. PhD Thesis, Normandie Université, École nationale d'ingénieurs de Sfax (Tunisia).

Ghenam, S., El Hami, A., Dammak, K., Gafsi, W., Akrout, A., Haddar, M. (2023a). Enhancing ball grid array (BGA) component design and reliability using a novel reliability-based design optimization (RBDO) methodology. *Mechanics of Advanced Materials and Structures*, 1–17. doi: 10.1080/15376494.2023.2237694.

Ghenam, S., El Hami, A., Gafsi, W., Akrout, A., Haddar, M. (2023b). Optimal performance and cost-effective design of BGA solder joints using a deterministic design optimization (DDO) under real operating conditions. *Microelectronics Reliability*, 146, 115019. doi: 10.1016/j.microrel.2023.115019.

Hadim, H.A. and Suwa, T. (2004). Integrated thermomechanical design and optimization of BGA packages using genetic algorithm. *ASMEDC*, 4, 495–502. doi: 10.1115/HT-FED2004-56512.

Halouani, A. (2020). Modélisation et simulation multi-physique pour la fiabilisation des composants electroniques. Doctoral Thesis, Université de Technologie de Troyes.

Kharmanda, G., Ibrahim, M.-H., Abo Al-kheer, A., Guerin, F., El-Hami, A. (2014). Reliability-based design optimization of shank chisel plough using optimum safety factor strategy. *Computers and Electronics in Agriculture*, 109, 162–171.

Lau, C., Abdullah, M.Z., Khor, C.Y. (2013). Optimization of the reflow soldering process with multiple quality characteristics in ball grid array packaging by using the grey-based Taguchi method. *Microelectronics International*, 30(3), 151–168. doi: 10.1108/MI-09-2012-0067.

Mabrouk, I.B., El Hami, A. (2019). Effect of number of blades on the dynamic behavior of a Darrieus turbine geared transmission system. *Mechanical Systems and Signal Processing*, 121, 562–578,

Makhloufi, A., Aoues, Y., El Hami, A. (2016). Reliability based design optimization of wire bonding in power microelectronic devices. *Microsystem Technologies*, 22(12), 2737–2748.

Moro, T., El Hami, A., El Moudni, A. (2002). Reliability analysis of a mechanical contact between deformable solids. *Probabilistic Engineering Mechanics*, 17(3), 227–232.

Müller, W.H. and Hauck, T. (2008). Simple methods for the durability assessment of microelectronic solders. *Mechanics of Advanced Materials and Structures*, 15(6–7), 485–498. doi: 10.1080/15376490802142437.

Sinda, G., El Hami, A., Wajih, G., Ali, A., Mohamed, H. (2024). A sensitivity study of BGA geometrical parameters based on electro-thermal fatigue analysis. In *Advances in Materials, Mechanics and Manufacturing*, Ben Amar, M., Ben Souf, M.A., Beyaoui, M. et al. (eds). Springer, Cham, 160–168.

Uday Kiran, K.A.V., Vigneshwar, N., Solomon, D.G., Arulraj, J.J. (2021). Significance of Taguchi design compared to response surface methodology in predicting optimal machining conditions for PCM process. *IOP Conference Series: Materials Science and Engineering*, 1128(1), 012019. doi: 10.1088/1757-899X/1128/1/012019.

Van Nhat, L. (2016) Modélisation de la tenue en fatigue des joints de brasure dans un module de puissance. Thesis, Université Paris-Saclay.

Zhang, J. and Zhang, D. (2013). Study of response surface methodology in thermal optimization design of multichip modules. *IEEE Transactions on Components, Packaging and Manufacturing Technology*, 3(12), 2075–2080. doi: 10.1109/TCPMT.2013.2271244.

4

Electronic Toll Collection in the Age of Connected Vehicles

Cooperative intelligent transportation systems (cooperative ITS or C-ITS) rely on the communications and sharing of information between vehicles, road infrastructures and communication infrastructure to improve users' safety, facilitate their trips and ensure the safety of road agents. This chapter presents the technological advances in the field of ITS in the context of the democratization of the communication between vehicle and infrastructure using ITS-G5. A new type of electronic toll collection service based on this type of communication is presented in this chapter. Its objective is to reduce maintenance costs, increase agents' safety and improve traffic fluidity.

4.1. Introduction

In recent years, intelligent transportation systems have attracted increasing interest from various players such as decision-makers, investors, companies, research laboratories, users, etc.

On the other hand, the European Commission published directive 2010/40/UE of the European Parliament and of the Council on July 7, 2010, concerning the context for the deployment of intelligent transportation systems in the field of road transportation and interfaces with other modes of transportation. It recalls that "The increase in the volume of road transport in the Union associated with the growth of the European economy and mobility requirements of citizens is the primary cause of increasing congestion of road infrastructure and rising energy consumption, as well as a source of environmental and social problems."

Chapter written by Adnane CABANI and Houcine CHAFOUK.

In the context of this directive, the fields that should be given priority include: 1) ITS applications to road safety and security, and 2) the connection between the vehicle and transport infrastructures. Therefore, road operators play a key role in the implementation of solutions addressing these increasing needs.

The communication between vehicle and infrastructure is important and can be used, among others, for vehicle geolocation. Thus, each vehicle becomes a valuable source of real-time information without any additional cost for the operators. The information gathered can be used to facilitate the payment in toll collection stations and improve traffic fluidity. This information can also be used to optimize routes and reduce the energy consumption of the vehicles, thus contributing to a reduction in the carbon footprint.

Since the beginning of 2018, Renault's Mégane model and PSA C4 and DS4 models have been equipped with a communication module (V2X) and marketed to the general public without any additional cost. In the context of the SCOOP@F project, over 3,000 intelligent vehicles will be deployed on 2,000 km of connected roads. Five deployment pilot sites in France are equipped with roadside units (RSU). These are Île-de-France, the East Corridor, the West, Bordeaux and Isère. The infrastructure is no longer an obstacle to the democratization of connected vehicles. It is in this context that we have proposed a new electronic toll collection service using ITS-G5 as part of CIFRE (Industrial Contract for Training through Research) (Randriamasy 2019).

4.2. Electronic toll collection service using ITS-G5

With the emergence of connected vehicles, it was important to propose a new electronic toll collection service that could be used to offer a seamless service to both connected and not connected vehicles. The proposed solution must coexist with the classical toll collection solution with a stop before the barrier, and with the DSRC (Dedicated Short Range Communications)-based solution.

The objective of this new service is to fluidify traffic by allowing connected vehicles to make the electronic payment without stopping (Randriamasy et al. 2019a, 2019b). When the operation is completed, the barrier rises and the vehicle passes without stopping. To reach this objective, we had to address the following issues:

– locate the vehicle with an accuracy of less than 1 meter in order to differentiate between two vehicles coming one after the other or several adjacent vehicles. This task must be accomplished while respecting GDPR (General Data Protection Regulation);

– secure communications and particularly electronic payments. Given the sensitivity of the exchanged data, any fraudulent use of the latter should be avoided;

– ensure data integrity, non-repudiation, confidentiality, authentication of the user, OBU (On-Board Unit) and RSU (Road Side Unit).

In our work, we have focused on the vehicle to infrastructure communication. The vehicles on the freeway will communicate through the on-board unit with the toll collection station equipped with a road side unit.

The advantages of this service for freeway operators are the following:

– reduction of maintenance costs: one RSU per toll plaza instead of one DSRC reader per lane;

– enhances agent safety: in the case of RSU, the intervention is conducted by the roadside, while for DSRC readers, the agent has to cross the lanes to be able to repair any broken down reader;

– better traffic fluidity: vehicles are not required to stop (payment by bank card or cash) or to slow down to 30 km/h (case of vehicles equipped with the DSRC badge).

In their articles, Randriamasy et al. (2019a, 2019b) use the CAM (Cooperative Awareness Message) messages sent by the connected vehicles to determine their position. Indeed, besides the GNSS (Global Navigation Satellite System) position, these messages contain information from the proprioceptive sensors of the vehicle. The roadside unit with a coverage of several hundred meters (being able to reach 1 km) is able to follow the path of vehicles as soon as they approach the toll collection station.

Our model (see Figure 4.1), known as the enhanced dynamic model (EDM), has various stages enabling the RSU to estimate as accurately as possible the position of vehicles.

The first stage involves the extraction of all useful data contained in CAM messages, namely message timestamping, GNSS coordinates, direction, velocity, acceleration, steering angle and turn rate.

The second stage involves data preprocessing. They must be expressed according to the International System of Units.

Moreover, according to the same principle of differential GPS, we calculate the difference between the actual position and the position indicated by the RSU station. We apply this gap to the vehicle positions provided via CAM messages.

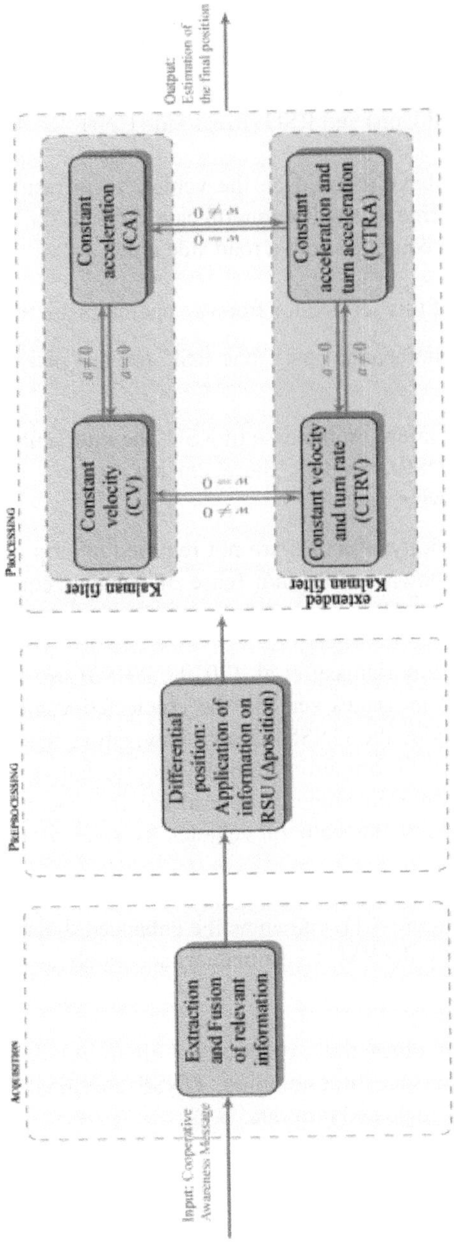

Figure 4.1. *Advanced dynamic model. For a color version of this figure, see www.iste.co.uk/elhami/applicationsofai.zip*

The last stage consists of the application of a Kalman filter or an extended Kalman filter for estimating the position of vehicles depending on their state. There are four possible states in which a vehicle can be every 100 ms:

– constant velocity, denoted by CV;

– constant acceleration (deceleration), denoted by CA;

– constant turn rate and velocity, denoted by CTRV;

– constant turn rate and acceleration, denoted by CTRA.

Figure 4.2 represents the changes in the state of parameters depending on the kinematic evolution of the vehicle.

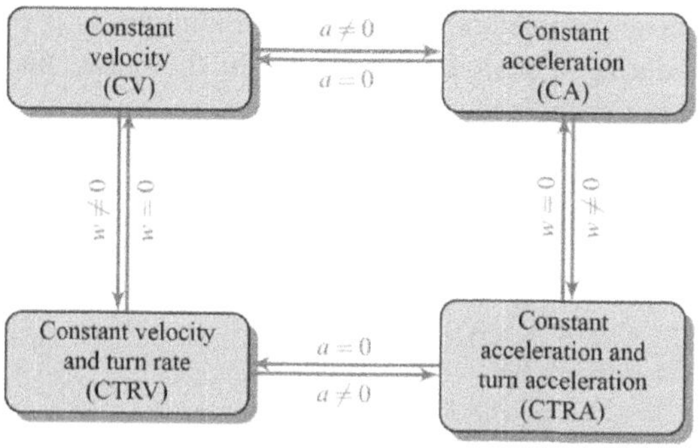

Figure 4.2. *Kinematic modeling of a vehicle*

For each state, we have defined a model. The details on the four models are provided in the following sections.

4.2.1. *Constant velocity (CV) model*

This model is used if the vehicle has CV. It applies the Kalman filter. We define our state vector as follows:

$$X_k = \left(x_k, y_k, v_{x_k}, v_{y_k}\right)$$

The equation that predicts the current state is defined by:

$$\hat{X}_k^- = A_k \hat{X}_{k-1}^+ + B_k u_k$$

This equation is reduced as follows:

$$\hat{X}_k^- = A_k \hat{X}_{k-1}^+$$

with the matrix A_k defined as follows:

$$A_k = \begin{pmatrix} 1 & 0 & dt_k & 0 \\ 0 & 1 & 0 & dt_k \\ 0 & 0 & 1 & 0 \\ 0 & 0 & 0 & 1 \end{pmatrix}$$

4.2.2. *Constant acceleration (CA) model*

This model, whose parameters have a linear evolution, is used in the case of a velocity variation during a period of time Δt. In this case, the Kalman filter is applied. We define our state vector as follows:

$$X_k = \left(x_k, y_k, v_{x_k}, v_{y_k}, a_{x_k}, a_{y_k}\right)$$

The current state prediction equation is defined by:

$$\hat{X}_k^- = A_k \hat{X}_{k-1}^+ + B_k u_k$$

This equation is reduced as follows:

$$\hat{X}_k^- = A_k \hat{X}_{k-1}^+$$

with the matrix A_k defined as follows:

$$A_k = \begin{pmatrix} 1 & 0 & dt_k & 0 & \frac{1}{2}dt_k^2 & 0 \\ 0 & 1 & 0 & dt_k & 0 & \frac{1}{2}dt_k^2 \\ 0 & 0 & 1 & 0 & dt_k & 0 \\ 0 & 0 & 0 & 1 & 0 & dt_k \\ 0 & 0 & 0 & 0 & 1 & 0 \\ 0 & 0 & 0 & 0 & 0 & 1 \end{pmatrix}$$

4.2.3. *Constant turn rate and velocity (CTRV) model*

This is a model whose parameters have a nonlinear evolution. This is why the extended Kalman filter is applied. The state vector is defined as follows:

$$X_k = (x_k, y_k, \theta_k, v_k, w_k)$$

The current state prediction equation is defined by:

$$\hat{X}_k^- = A_k \hat{X}_{k-1}^+ + B_k u_k$$

This equation is reduced as follows:

$$\hat{X}_k^- = A_k \hat{X}_{k-1}^+$$

with the matrix A_k defined as follows:

$$\hat{X}_k^- = \hat{X}_{k-1}^+ + \begin{pmatrix} \frac{v}{w}\sin(wdt_k+\theta(t))-\frac{v}{w}\sin(\theta(t)) \\ \frac{v}{w}\cos(wdt_k+\theta(t))-\frac{v}{w}\sin(\theta(t)) \\ wdt_k \\ 0 \\ 0 \end{pmatrix}$$

where θ is the direction of the vehicle. Figure 4.3 presents a simple block diagram showing the characteristics of the vehicle.

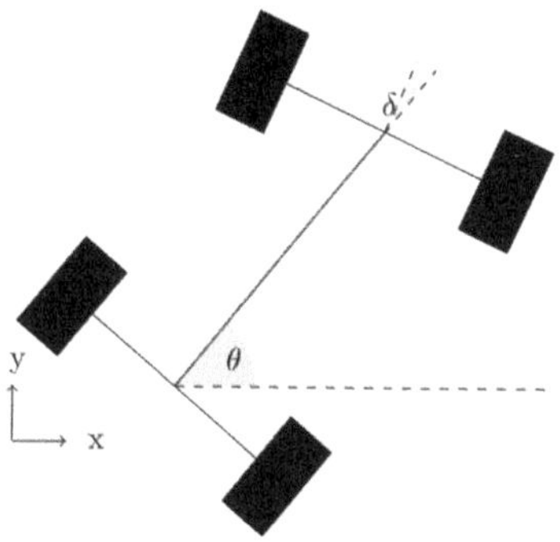

Figure 4.3. *Characteristics of the vehicle. For a color version of this figure, see www.iste.co.uk/elhami/applicationsofai.zip*

4.2.4. *Constant turn rate and acceleration (CTRA) model*

This is a model whose parameters have a nonlinear evolution. Therefore, if the vehicle is in this state, the extended Kalman filter will be applied. We define our state vector as follows:

$$X_k = (x_k, y_k, \theta_k, v_k, a_k, w_k)$$

The current state prediction equation is:

$$\hat{X}_k^- = A_k \hat{X}_{k-1}^+$$

with the matrix A_k defined as follows:

$$\hat{X}_k^- = \hat{X}_{k-1}^+ + \begin{pmatrix} \Delta x \\ \Delta y \\ wdt_k \\ adt_k \\ 0 \\ 0 \end{pmatrix}$$

where

$$\Delta x = \frac{1}{w^2}\Big(\big(v(t)w + awdt_k\big)\sin\big(wdt_k + \theta(t)\big) + a\cos\big(wdt_k + \theta(t)\big) - v(t)w\sin(\theta(t) - a\cos(\theta(t))\Big)$$

$$\Delta y = \frac{1}{w^2}\big(\big(-v(t)w + awdt_k\big)\cos\big(wdt_k + \theta(t)\big) + a\sin(wdt_k + \theta(t)) + v(t)w\cos(\theta(t) - a\sin(\theta(t))\big)$$

4.3. Experimental results of the path tracking

In order to test our EDM model (Randriamasy et al. 2017, 2018), we used the following equipment:

- one Trimble GPS (portable Trimble Geoexplorer GeoXH 6000 series system);
- one Neavia V2X communication module;
- one instrumented vehicle;
- one roadside unit.

Several routes were taken. This section presents the results obtained on the Taissy site (direction Province – Paris). The vehicle path is shown in Figure 4.4.

The parameter evolutions trigger one of the models (CV, CA, CTRV and CTRA) to estimate the position of the vehicle. Figure 4.5 presents the results obtained.

It can be noted that our EDM method yields better results compared, on the one hand, with any other model, and, on the other hand, to the values provided by the GPS sensor. The mean square error of our EDM model is 0.15 m.

Figure 4.4. *Path of the vehicle – Taissy site. For a color version of this figure, see www.iste.co.uk/elhami/applicationsofai.zip*

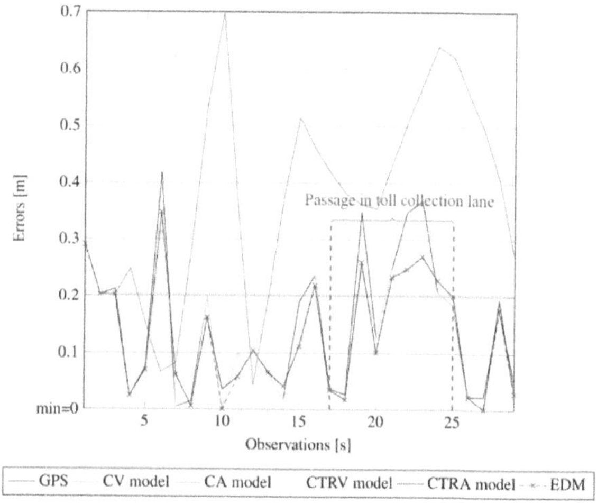

Figure 4.5. *Gaps between CV, CA, CTRV, CTRA, EDM model filters and GPS and the reference path – Taissy site (direction Province – Paris). For a color version of this figure, see www.iste.co.uk/elhami/applicationsofai.zip*

The final objective of this work is to be extended to the application of a free flow electronic toll collection as illustrated in Figure 4.6 in order to:

– eliminate congestion in toll collection stations;

– fluidify traffic;

– reduce particle (carbon, NOx, etc.) emission.

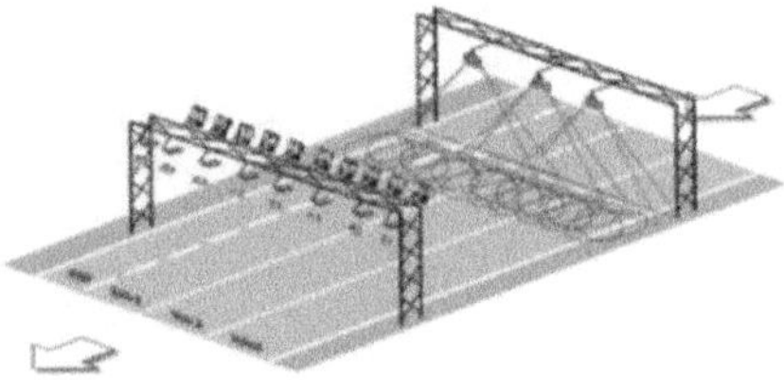

Figure 4.6. *Free flow toll collection (ETSI 2020). For a color version of this figure, see www.iste.co.uk/elhami/applicationsofai.zip*

4.4. Conclusion

The perspective of a world populated by connected and/or autonomous cars offers countless possibilities due to the expected future services, both as a driver and as a service provider.

With this new service, we have brought a new geolocation solution with an accuracy of less than 1 m from vehicles connected by the ITS-G5 technology. User anonymity is preserved. Only the competent authorities can establish the link between the service user and the electronic payment.

The expected objective is to implement a standard according to which various highway operators can use the service. A pre-standard was launched in September 2020 under the name ETSI TR 103 579 (ETSI 2020).

The proposed solution may serve for other types of services and other problem-oriented software particularly for geolocation on special lanes, contraflow detection, guiding of autonomous robots, etc.

4.5. References

ETSI (2020). Intelligent Transport Systems (ITS); Pre-Standardization Study on payment applications in Cooperative ITS using V2I communication. Technical Report [Online]. Available at: https://portal.etsi.org/webapp/WorkProgram/Report_WorkItem.asp?wki_id=53992 [Accessed 4 April 2021].

Randriamasy, M. (2019). Localisation et transmissions sécurisées pour la communication Véhicule à Infrastructure (V2I) : application au service de télépéage ITS-G5. PhD Thesis, Normandie Université [Online]. Available at: https://tel.archives-ouvertes.fr/tel-02148434 [Accessed on 4 April 2021].

Randriamasy, M., Cabani, A., Chafouk, H., Fremont, G. (2017). Reliable vehicle location in electronic toll collection service with cooperative intelligent transportation systems. *2017 IEEE 28th Annual International Symposium on Personal, Indoor, and Mobile Radio Communications (PIMRC)*, 1–7.

Randriamasy, M., Cabani, A., Chafouk, H., Fremont, G. (2018). Evaluation of methods to estimate vehicle location in Electronic Toll Collection Service with C-ITS. *2018 IEEE Intelligent Vehicles Symposium (IV)*, 748–753.

Randriamasy, M., Cabani, A., Chafouk, H., Fremont, G. (2019a). Formally validated of novel tolling service with the ITS-G5. *IEEE Access*, 7, 41133–41144.

Randriamasy, M., Cabani, A., Chafouk, H., Fremont, G. (2019b). Geolocation process to perform the electronic toll collection using the ITS-G5 technology. *IEEE Transactions on Vehicular Technology*, 68(9), 8570–8582.

5

At the Core of Artificial Intelligence: Leveraging Machine Learning through Random Forest

Artificial intelligence (AI) has captivated the imagination of mankind for decades, and its influence on our lives is steadily growing. This technological revolution has machine learning at its core, a fundamental branch of AI enabling machines to learn from data and make autonomous decisions.

The random forest is an innovative approach in the field of machine learning, offering a robust and multipurpose method for classification and regression. The exploration in this chapter presents the theoretical foundations of this technique, its practical applications and the challenges associated with its deployment.

By combining an in-depth comprehension of AI with a detailed examination of random forest, we are ready to venture into a field where human–machine boundary is blurred, opening the way to new perspectives and wide possibilities. The reader is invited to prepare for diving into the fascinating world of AI and machine learning, where imagination joins reality to shape the future of our society.

The next sections of this chapter will explore in detail the basics of decision tree construction and also the process of random forest creation. We will examine the underlying principles that guide these machine learning methods and will dive into their operation. Furthermore, we will approach the resolution of a malware detection problem in mobile applications using the random forest algorithm. We will study how this approach can be successfully applied to identify and prevent potential threats in the field of computer security.

Chapter written by Anas EL ATTAOUI, Younes KOULOU and Norelislam EL HAMI.

5.1. Overview of AI and its branches

AI is a fascinating research domain that aims to provide machines with intellectual abilities similar to those of human beings. This constantly evolving discipline explores various techniques and approaches that enable computers to conduct complex tasks, such as voice recognition, computer vision, decision-making and many more.

AI became an essential pillar in many scientific domains such as big data (Najim et al. 2014; El Hami et al. 2017a, 2017b; El Attaoui et al. 2023), data analysis and mathematical optimization (Koulou et al. 2020; Bouchekourte and El Hami 2022; Rhouas and El Hami 2022). Using the capacities of AI, researchers and practitioners are able to explore, understand and use massive datasets in a previously unimaginable manner. Machine learning algorithms, for example, allow AI systems to identify complex models and make decisions based on these models, a particularly useful tool in domains such as prediction, classification and recommendation. Similarly, mathematical optimization (Zeine et al. 2017; Zemzami 2019; Zemzami et al. 2019; El Attaoui et al. 2023; Rhouas et al. 2022), which aims to find the best possible solutions to complex problems, benefits from the integration of AI techniques. Deep learning methods, in particular, have shown their efficiency in solving difficult optimization problems (El Hami et al. 2015a, 2015b, 2017a, 2017b; Zemzami 2017; Zemzami et al. 2020), by allowing AI systems to rapidly and efficiently explore the space of solutions. By combining AI with data analysis and mathematical optimization, researchers can transform raw data into usable information and discover optimal solutions to otherwise overwhelming problems (El Hami et al. 2010, 2011, 2013, 2015a, 2015b).

5.1.1. *AI and machine learning*

A significant number of advances in the field of AI rely on machine learning, an essential branch of this field. Machine learning involves the development of algorithms and computer models that can learn from data. Unlike standard programs that follow strict instructions defined by programmers, machine learning systems can analyze data, detect diagrams and consequently adjust their behavior, while adapting to new information.

AI and machine learning are closely interrelated. Indeed, machine learning is often the underlying driver that powers many AI applications. By using it, AI systems enhance their performance in time, by relying on experience and available data.

5.1.2. *Robust techniques are required for classification and regression*

In the field of machine learning, classification and regression are two fundamental tasks. Classification consists of assigning labels or categories to databases on their characteristics, while regression aims to predict a numerical value based on input variables (Castillo-Botón et al. 2022).

The quality of predictions produced by machine learning algorithms depends, to a large extent, on the robustness of these algorithms. Robust techniques must be able to effectively process sets of complex, noised or incomplete data, while avoiding overlearning, when the model is too closely adapted to training data and does not generalize properly when fed new data (Belagoune et al. 2021).

5.1.3. *Introduction to the random forest method*

The random forest method is one of the most popular and efficient techniques used in machine learning for classification and regression (Mojsilović et al. 2023). This method relies on the use of a set of decision trees, each tree being randomly constructed based on a subset of training data (Zhang et al. 2024).

By combining the predictions of multiple decision trees, the random forest manages to reduce overlearning and enhance the robustness of the global model. Moreover, it offers high flexibility and the capacity to manage complex data, which explains its frequent selection for many applications in fields such as medicine, finance, biology and many others (Hasan and Horvat 2024).

5.2. Decision trees

The random forest algorithm is mainly based on decision trees. In order to fully grasp its operation, it is essential to understand the mechanism of decision trees. This section explores in detail the operation of decision trees, as well as various types of algorithms underlying them, their advantages and their limits.

5.2.1. *Structure of a decision tree and associated terms*

Decision trees are widely used in machine learning, their operation relying on heuristics developed with supervised learning techniques. These are hierarchical structures, composed of nodes and leaves connected by branches. In their graphic representation, the root is in the upper part and the leaves are in the lower part. The internal nodes, also known as decision nodes, may contain one or more rules, also known as tests or conditions (Wang et al. 2020).

The various values that a variable may take in a decision tree are referred to as instances or attributes. The end nodes contain the class to predict, also known as target variable. After its construction, a decision tree can be expressed in the form of a set of decision rules. When the variable to be predicted is qualitative, the tree takes the form of a classification tree. On the other hand, in the case of a quantitative variable, it is a regression tree. Figure 5.1 represents the structure of a classification decision tree.

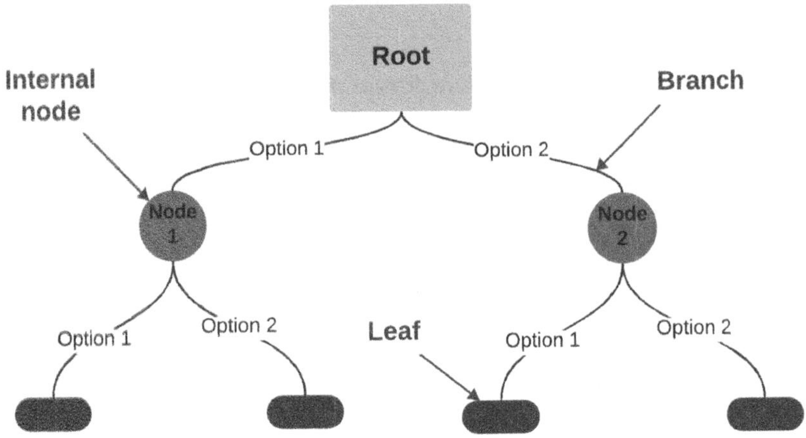

Figure 5.1. *Structure of a decision tree. For a color version of this figure, see www.iste.co.uk/elhami/applicationsofai.zip*

5.2.1.1. *Example: cardiac disease diagnosis*

Let us assume that the structure of the decision tree for cardiac disease diagnosis is as follows:

– if the patient has a chest pain and an abnormal electrocardiogram, then predict **"cardiac disease"**;

– otherwise, if the patient has a chest pain and high cholesterol level, then predict **"cardiac disease"**;

– otherwise, predict "**no cardiac disease**".

In the above-mentioned example, the decision tree illustrated in Figure 5.2 takes into account two main criteria: the presence of chest pain and the results of medical tests. If the patient meets one of these defined conditions, the tree predicts that they have a heart disease, identified by "**yes**". Otherwise, it predicts the absence of heart disease, denoted by "**no**".

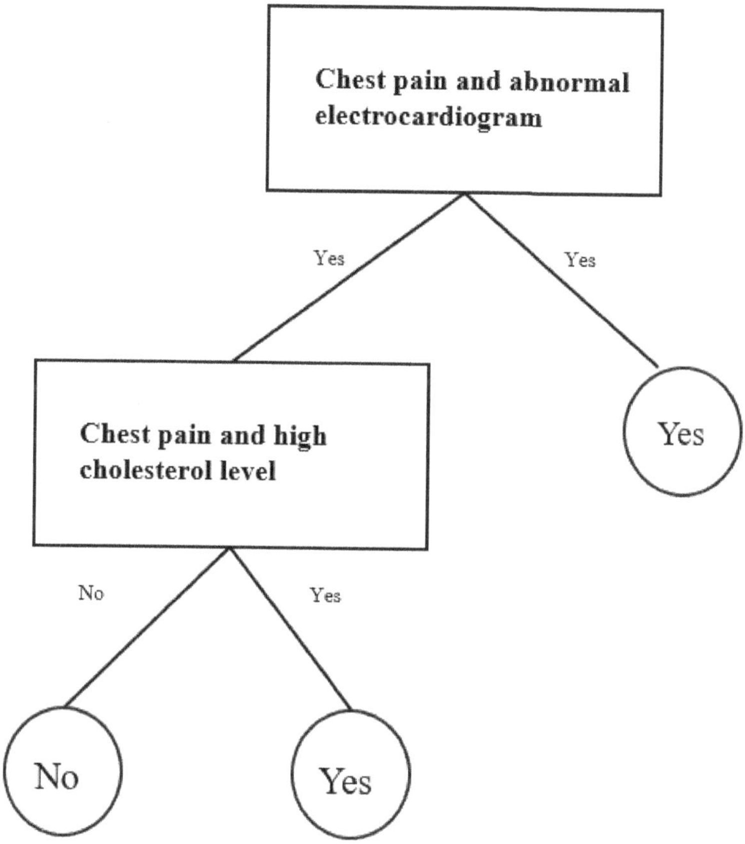

Figure 5.2. *Structure of a decision tree*

5.2.1.2. *Generic algorithm for building a decision tree.*

The structure of the algorithm illustrated in Figure 5.3 allows it to initialize the decision tree, and then loop through each node until the tree is entirely constructed. At each loop iteration, it verifies if the current node is an end node. If this is the case, it assigns a class to this node. If not, it selects a test criterion and creates branches corresponding to each response possibility of the test. Finally, it advances to the following node in the tree along a predefined path, until the tree is completed.

```
Begin
Define tree as an empty tree
Define the current node as the tree root
While the decision tree is not complete Do
    If the current node is an end node Then
        Assign a class to the current node
    Else
        Select a test criterion
        For each test response possibility Do
            Create a new branch with this response
        End For
    End If
    Go to the next node along a predefined path
End While
End
```

Figure 5.3. *Generic algorithm for decision tree construction*

5.2.2. *Entropy*

Let us consider a probability distribution P represented by $(p_1, p_2, \ldots, p_n)$. The entropy $E(P)$ of this distribution is the measure of the information it contains, as specified in equation [5.1]:

$$E(P) = -\sum_{k=1}^{n} pi * \log pi \qquad [5.1]$$

where log represents the binary logarithm function.

Let us consider a dataset T characterized by n classes $(C_1, C_2, \ldots, C_n)$ according to the target variable. The amount of information required to determine the class of an individual of the set T corresponds to the entropy $E(P)$, where P is the probability distribution of the partition $(C_1, C_2, \ldots, C_n)$:

$$P = \left(\frac{|C1|}{|T|}, \frac{|C2|}{|T|}, ..., \frac{|Cn|}{|T|} \right) \quad [5.2]$$

The entropy of *T* will therefore be:

$$E(T) = -\sum_{k=1}^{n} \frac{|C1|}{|T|} * \log \frac{|C1|}{|T|} \quad [5.3]$$

If the dataset T is partitioned into subsets (T_1, T_2, …, T_m), where m is the number of values that the considered attribute X can have, the amount of information required to identify the class of an individual belonging to T_i can be expressed as follows:

$$E(X,T) = \sum_{j=1}^{m} \frac{|Tj|}{|T|} * E(Tj) \quad [5.4]$$

5.2.3. *Information gain*

Let us consider a dataset *T*. The information gain of T with respect to a given partition T_j can be defined as the difference in entropy resulting from the partition of *T* along T_j. In other terms, it measures the reduction of uncertainty on data *T* when they are divided into subsets as a function of the partition T_j

$$Gain(X,T) = E(T) - E(X,T) = E(T) - \sum_{j=1}^{m} \frac{|Tj|}{|T|} * E(Tj) \quad [5.5]$$

5.2.4. *Gini index*

The Gini index is a measure of impurity used in decision trees to evaluate the quality of a dataset division. It is calculated by assessing the probability of a randomly chosen element in a subset to be misclassified according to the distribution of classes in this subset. More precisely, in the context of decision trees, the Gini index measures to what point a node is impure by considering the distribution of classes in this node. A Gini index near zero indicates high purity, which means that the node contains mainly elements of a single class, while a Gini index near one indicates a higher impurity, which means that the node contains an equal or nearly equal distribution of various classes. By choosing the division that minimizes the Gini index, the decision trees aim to create partitions that maximize the purity of child nodes

$$Gini(T) = 1 - \sum_{j=1}^{m} \left(\frac{|Tj|}{|T|} \right)^2 \qquad [5.6]$$

As the dataset T is segmented into subsets (T_1, T_2, …, T_m) as a function of the variable X:

$$Gini\ index(X,T) = \sum_{j=1}^{m} \frac{|Tj|}{|T|} * Gini(Tj) \qquad [5.7]$$

5.2.5. *Decision tree generation algorithm*

There are several approaches available for decision tree construction, thus offering the possibility to select among various variables. In this section, we present the principle of three algorithms for decision tree construction: ID3, C4.5 and CART.

5.2.5.1. *Decision tree construction algorithm*

Figure 5.4 shows the process of decision tree construction from a set of samples and a set of attributes. The objective of the decision tree is to classify data according to the values of these attributes.

The algorithm begins by verifying whether all samples belong to the same class. If this is the case, it creates a leaf in the tree labeling it by this class and returns the constructed tree. If not, it continues by searching for the best attribute to be used to divide data into subsets.

The process is then repeated for each possible value of this attribute, by creating a corresponding node and subdividing the samples depending on this value. For each resulting subset, the algorithm recursively recalls the tree construction function until a termination condition is encountered. This condition can be defined depending on various criteria, such as the maximal depth of the tree, the minimum number of samples in a node or other measures of data impurity.

The algorithm continues to create nodes and divide data until the termination condition is met for all of the paths of the tree. Once terminated, it returns the constructed decision tree, which can be used to classify new samples depending on their attributes.

Begin

Inputs: E (Set of samples), L (Set of attributes)

Output : a decision tree

1. **If** all the samples in E belong to the same class **then**
2. Create and return a leaf labelled by this class
3. **Else**
4. **Repeat**
5. Find the best attribute in the set L: A
6. Separate the samples of E into n subsets, according to the value of this attribute
7. Allocate a node labelled by A
8. **For** each value v of A do
9. **Repeat** Decision_Tree_Construction (subset corresponding to v, L \{A} **until** termination condition
10. Add the resulting child node to that labelled by A
11. **Until** the termination condition

End

Figure 5.4. *Decision tree construction algorithm*

ID3, C4.5 and CART algorithms are essential in the construction of decision trees, sharing a fundamental operating principle. Their central mechanism consists of creating a tree structure based on the analysis and classification of attributes. However, what differentiates them is their approach related to the type of variables used and the manner in which these attributes are classified within the tree.

5.2.5.2. *ID3 algorithm*

ID3 recursive algorithm operates by iteration through a set of attributes, a set of target attributes and a set of samples, being limited to qualitative attributes. At each stage, it chooses the attribute that maximizes the information gain, thus enabling an efficient classification of data at this tree level. Once selected, this attribute is used to divide data into subsets, thus contributing to the progressive construction of the decision tree. This approach, focused on the maximization of information gain at each stage, aims to create an optimal decision tree for the classification of categorical data (Chanmee and Kesorn 2023).

5.2.5.3. *CART algorithm*

The recursive CART (Classification And Regression Trees) algorithm is a machine learning method that enables the construction of decision trees by maximizing the Gini index. It is an algorithm that generates binary decision trees, where each node of the tree represents a decision based on the value of a particular attribute. CART is broadly used in the fields of classification and regression, offering an efficient approach to data partition into homogeneous subsets in terms of target variable (Zhao et al. 2024).

5.2.5.4. *C4.5 algorithm*

ID3 algorithm, initially designed to manage only qualitative attributes, was improved to include the management of quantitative attributes, thus leading to the C4.5 algorithm. The latter uses the entropy gain function for selecting the attributes to be considered at each level of the decision tree, thus contributing to a better adaptation to numerical data. Consequently, the C4.5 algorithm was further improved and renamed C5, offering a faster execution and a more efficient memory management. The decision trees obtained with C5 also have an important characteristic: they reduce decision redundancy, which contributes to a better comprehension of models and to optimized performances in terms of memory and execution time (Li et al. 2024).

5.2.6. *Example of decision tree construction*

In Table 4.1, each attribute has the following options: small, medium or large for size, light or heavy for weight, and short or long for hair length. The class to be defined is the breed, which can be either dog or cat.

Size	Weight	Hair length	Breed (dog/cat)
Small	light	short	Cat
Medium	light	short	Cat
Medium	heavy	short	Dog
Large	light	long	Dog
Large	heavy	long	Dog
Medium	light	long	Cat
Large	light	short	Cat

Table 5.1. *Data table*

This data will be used to build a decision tree using the ID3 algorithm. The first step is to calculate the initial entropy, and then the information gain for each characteristic. This will be followed by the selection of the characteristic that maximizes the information gain for the first tree division. This process will be repeated until obtaining the final tree.

5.2.6.1. *Entropy calculation*

$$E(breed)?$$

The breed contains four cats and three dogs, and then based on equation [5.3], the entropy is:

$$E(breed) = -\frac{4}{7} * \log\frac{4}{7} - \frac{3}{7} * \log\frac{3}{7}$$

$$*E(breed) = 0.985$$

$$E(breed, size)?$$

The size is characterized by three options: "small", "medium" and "large". To calculate the size entropy, it is first necessary to calculate the entropies for small, medium and large.

For "small", there is one cat and no dogs. Therefore, the entropy is 0, as all of the examples belong to a single class

$$E(small) = 0$$

For "medium", there are two cats and one dog:

$$E(medium) = E(2\ cats,\ 1\ dog) = -\frac{2}{3} * \log\frac{2}{3} - \frac{1}{3} * \log\frac{1}{3}$$

$$E(medium) = 0.918$$

For "large", there is one cat and two dogs

$$E(large) = E(1\ cat,\ 2\ dogs) = -\frac{1}{3} * \log\frac{1}{3} - \frac{2}{3} * \log\frac{2}{3}$$

$$E(large) = 0.918$$

According to equation [5.4], the "size" entropy taking into account the target variable "breed" $E(breed, size)$ is:

$$E(breed, size) = -\frac{1}{7} * E(small) + \frac{3}{7} * E(medium) + \frac{3}{7} ** E(large)$$

$$*E(breed, size) = 0.787$$

$$E(breed, weight)?$$

The weight is characterized by two options: "light" and "heavy". Similarly, in order to calculate the weight entropy, we first calculate the entropy for light and heavy.

For "light", there are four cats and one dog:

$$E(light) = E(4\ cats,\ 1\ dog) = -\frac{4}{5} * \log\frac{4}{5} - \frac{1}{5} * \log\frac{1}{5}$$

$$E(light) = 0.721$$

For "heavy", there is one cat and one dog. Therefore, the entropy will be maximal and equal to one, as we have an equitable distribution of classes

$$E(heavy) = 1$$

hence:

$$E(breed, weight) = \frac{5}{7} * E(light) + \frac{2}{7} * E(heavy)$$

$$*E(breed, weight) = 0.800$$

$$E(breed, hair\ length)?$$

Hair length is characterized by two options: "short" and "long".

For "short", there are three cats and one dog

$$E(short) = E(3\ cats,\ 1\ dog) = -\frac{3}{4} * \log\frac{3}{4} - \frac{1}{4} * \log\frac{1}{4}$$

$$E(short) = 0.811$$

For "long", there is one cat and two dogs

$$E(long) = E(1\ cat,\ 2\ dogs) = -\frac{1}{3} * \log\frac{1}{3} - \frac{2}{3} * \log\frac{2}{3}$$

$$E(long) = 0.918$$

$$E(breed, hair\ length) = \frac{4}{7} * E(short) + \frac{3}{7} * E(long)$$

$$*E(breed, hair\ length) = 0.856$$

5.2.6.2. *Gain calculation*

After having calculated the entropy, it is time to calculate the information gain and to determine the attribute that maximizes this gain.

According to equation [5.5], the information gains of "size", "weight" and "hair length" attributes are:

$$Gain(breed, size) = E(breed) - E(breed, size)$$
$$Gain(breed, size) = 0.985 - 0.787$$
$$*Gain(breed, size) = 0.198$$
$$Gain(breed, weight) = E(breed) - E(breed, weight)$$
$$Gain(breed, weight) = 0.985 - 0.800$$
$$*Gain(breed, weight) = 0.185$$
$$Gain(breed, hair\ length) = E(breed) - E(breed, hair\ length)$$
$$Gain(breed, hair\ length) = 0.985 - 0.856$$
$$*Gain(breed, hair\ length) = 0.129$$

The attribute that maximizes information gain is the "size", which means that it will be placed at the top of the decision tree, as represented in Figure 5.5.

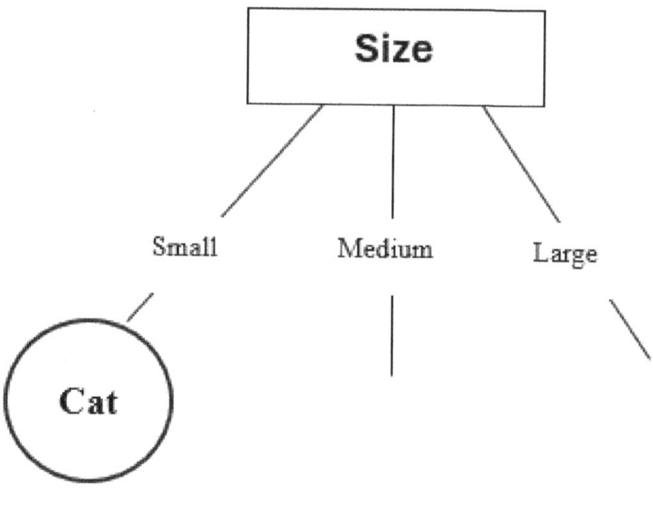

Figure 5.5. *The top of the decision tree*

In order to define the attributes connected to the branches with "size = medium" and "size = large" conditions, we must filter our data on "size = medium" and recalculate entropy and gain. This operation will also be repeated for "size = large". Table 5.2 presents the data subset that verifies "size = medium" condition.

Size	**Weight**	**Hair length**	**Breed (dog/cat)**
Medium	Light	Short	Cat
Medium	Light	Short	Dog
Medium	Light	Long	Cat

Table 5.2. *Subset of "size = medium" data*

$$E(breed)?$$

$$E(breed) = E(2\ cats,\ 1\ dog) = -\frac{2}{3} * \log\frac{2}{3} - \frac{1}{3} * \log\frac{1}{3}$$

$$*E(breed) = 0.918$$

$$E(breed, weight)?$$

Weight is always characterized by two options "light" and "heavy".

For "light", there are two cats and 0 dogs. Therefore, the entropy is 0, as all of the examples belong to a single class.

$$E(light) = 0$$

For "heavy", there are 0 cats and one dog:

$$E(heavy) = 0$$

hence:

$$*E(breed, weight) = 0$$

$$E(breed, hair\ length)?$$

Hair length is characterized by "short" and "long".

For "short", there is one cat and one dog. Then,

$$E(short) = 0.5$$

For "long", there is one cat and 0 dogs. Therefore,

$$E(long) = 0$$

Therefore:

$$E(breed, hair\ length) = \frac{2}{3} * E(short) + \frac{1}{3} * E(long) = \frac{2}{3} * 0.5 + \frac{1}{3} * 0$$

$$* E(breed, hair\ length) = 0.333$$

And therefore, the information gains are:

$$Gain\ breed, weight \quad 0.918$$

$$Gain\ breed,\ hair\ length \quad 0.585$$

The attribute that maximizes information gain is weight. The continuity of the decision tree is thus illustrated in Figure 5.6.

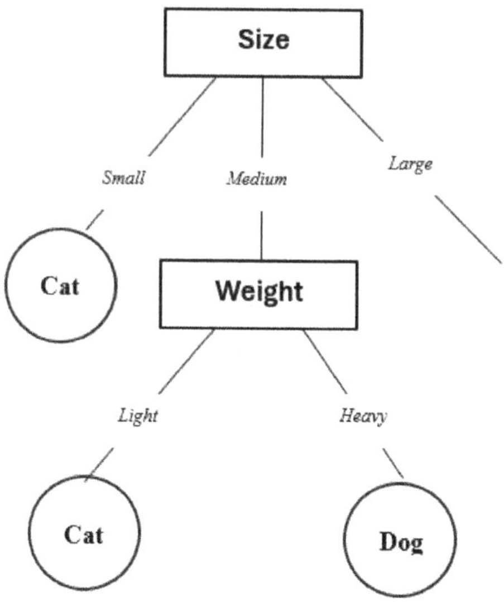

Figure 5.6. *Decision tree with "size=medium"*

Applying the same principle for "size = large", we obtain the decision tree in Figure 5.7.

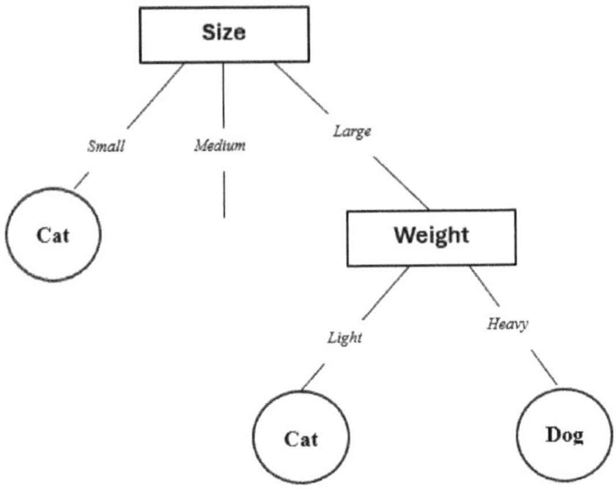

Figure 5.7. *Decision tree with "size = large"*

Figure 5.7 presents the final decision tree.

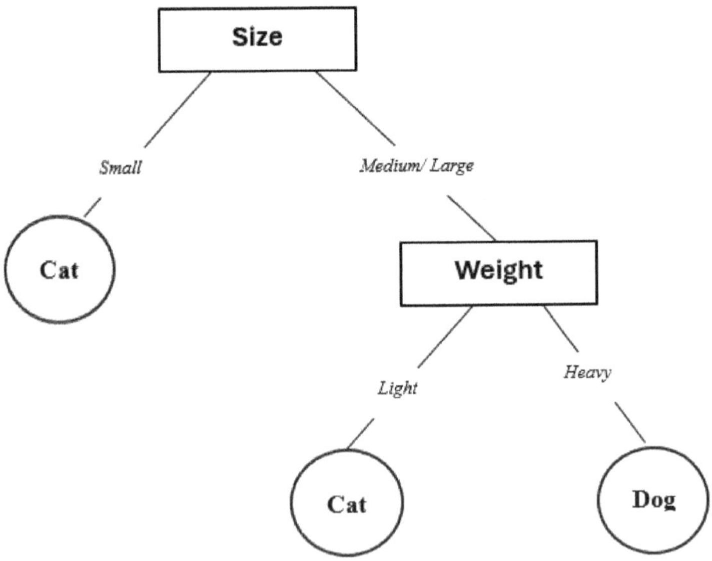

Figure 5.8. *Final decision tree*

5.2.7. *Advantages and limitations of decision trees*

Decision trees are widely used in machine learning due to their numerous advantages. First, they can be easily interpreted and visualized. The tree hierarchical structure can be used for an intuitive representation of decisions made at each stage, which renders decision trees accessible even to nonexperts. Moreover, decision trees are able to efficiently manage mixed data, datasets involving both categorical and continuous variables, therefore a multipurpose choice for various classification and regression tasks.

Another advantage of decision trees is their capacity to manage complex interactions between variables. Unlike certain other machine learning methods, such as linear models, decision trees are able to capture nonlinear interactions between the characteristics, being thus able to adapt to situations in which the relations between variables are complex or nonlinear.

Nevertheless, despite their many advantages, decision trees have significant limitations. A major obstacle is their overfitting tendency, which is particularly strong for deep and complex trees. Indeed, a too complex tree risks memorizing the training data instead of generalizing to new data, leading to average performances on unobserved datasets. To address this problem, regularization techniques such as tree-depth restriction or the reduction in the number of leaves can be implemented. Moreover, decision trees are often sensitive to minor input data variations, which may induce instability in their structure and fluctuating performances. Furthermore, the construction of optimal decision trees may prove time consuming, particularly with large datasets or complex trees. The random forest tree was mainly developed to overcome these limitations.

In conclusion, decision trees offer an intuitive and multipurpose method for classification and regression, but they should be carefully used to avoid overfitting and take into account the computational costs associated with their construction.

5.3. Random forest foundations

5.3.1. *Origin and basic principle of the random forest*

The random forest algorithm was introduced by Leo Breiman and Adele Cutler in the 2000s. It is based on previous research on decision trees and overall methods. Breiman proposed this algorithm as a robust alternative to individual decision trees, which may be sensitive to overfitting and to data variability (Zhou et al. 2023). Since its introduction, the random forest became one of the most popular supervised learning algorithms due to its simplicity, robustness and to its capacity to efficiently manage large-size and complex datasets (Wan et al. 2023).

The random forest is a supervised learning algorithm used for classification, regression and other tasks related to predictive analysis. Its basic principle relies on combining several individual decision trees to form a robust and diversified “forest”. Each decision tree is formed on a random sample of training data and uses a random subset of characteristics available for making decisions. When the time comes to make predictions, each tree of the forest gives its own prediction and the final prediction is determined by majority vote or average, depending on whether it is a classification or regression task (Antoniadis et al. 2021).

The random forest has several advantages compared to other supervised learning algorithms, such as individual decision trees, linear regression methods and support vector machines (SVM). First, it is less susceptible of overfitting training data, as it combines the predictions of several decision trees. Furthermore, it can efficiently manage datasets with highly correlated characteristics and missing values. Finally, the random forest is also capable of providing estimations on the importance of variables, which may be useful for interpreting the models and selecting the most informative characteristics (Toft et al. 2023).

5.3.2. *Stages of random forest algorithm*

The random forest algorithm starts by creating a large number of random decision trees. Each tree is formed of a random subset of training data, also by randomly selecting a subset of characteristics at each decision node. Then, each tree provides its own prediction for a new observation, either a classification or a regression. The predictions of all trees are then combined, either by vote for classification, or by mean for regression, in order to produce the final prediction. This approach can be used to reduce overfitting and provides a measure of the importance of characteristics to help interpreting the results (Yan et al. 2023).

5.3.2.1. *Creation and training of decision trees*

The first stage of the random forest algorithm involves the creation of a large number of decision trees. Each tree is randomly elaborated by selecting a random sample of training data and a random subset of characteristics at each decision point. Each tree is formed on a random sample of training data (Bootstrap), in order to segment the data according to characteristics in order to optimize the purity of the tree nodes. The commonly used measure to evaluate this purity is the Gini index for the classification and the reduction of the mean square error for the regression. This approach generates a diversity of trees, ensuring a robust and generally applicable data representation.

5.3.2.2. *Combination of predictions*

Once all trees are constructed, when it is time to make a prediction for a new observation, each tree of the forest gives its own prediction. For a classification, each tree votes for the most frequent class, while for a regression, the predictions of all trees are averaged to give the final prediction.

When a new instance must be classified or predicted, it goes through each tree of the forest, and the final prediction is obtained according to the previously described rule.

5.3.2.3. *Combination of predictions*

The fusion of predictions issued from several trees contributes to attenuating the overfitting compared to a single decision tree, given that individual trees often have the tendency to overfit training data. This aggregation of predictions coming from many trees renders the global model more robust and generally applicable. Furthermore, a significant advantage of the random forest is its capacity to assess the importance of characteristics. This measure can be used to identify the characteristics that have the highest influence on the predictions of the model, thus offering valuable information for the comprehension and interpretation of results.

The stages described above are presented in the form of algorithm in Figure 5.8.

```
Begin
Inputs: E (Set of samples), Tree_Number, Max_Depth, Min_Leaf_Size
Output: A random forest
For each Tree from 1 to Tree_Number do
    Create a random Bootstrap sample from E
    Select a random subset of characteristics from the total set of characteristics
    Build a decision tree with the bootstrap sample and the subset of characteristics
  until the maximal depth or the minimal size of the leaf is reached
For each Observation in E do
    Tree_Predictions = []
    For each tree in the Random Forest do
        If Task is Classification, then
            add Tree Class Prediction to Tree_Predictions
        Else if Task is Regression, then
            Add Tree Value Prediction to Tree_Predictions
    If Task is Classification, then
        Final_Prediction = Majority_Vote (Tree_Predictions)
    Else if Task is Regression, then
        Final_Prediction = Mean (Tree_Predictions)
    Display Final_Prediction
End
```

Figure 5.9. *Random forest algorithm*

5.3.3. *Random forest example for classification*

To facilitate comprehension, we decided to apply the random forest algorithm to the data of Table 5.3.

This approach is used to illustrate the algorithm operation step by step. Using available data, the random forest algorithm will construct a set of many decision trees, each being formed on a random subset of training data and available attributes. At each stage of the tree construction, the algorithm randomly selects a subset of attributes to divide the data, thus trying to maximize the purity of the resulting subgroups.

This process is repeated for each tree of the forest, and once all the trees have been constructed, they collectively vote to determine the most probable class for each data instance.

Table 5.3 presents six data instances, characterized by four distinct attributes denoted by x_0, x_1, x_2 and x_3, identified by id. Each instance is associated with a determined class, represented by the output variable y, taking binary values, either 0 or 1.

id	x_0	x_1	x_2	x_3	y
0	4.4	2.9	1.4	0.2	0
1	5.3	3.7	1.5	0.2	0
2	5.4	3.9	1.7	0.4	0
3	6.6	3.0	4.4	1.4	1
4	5.5	2.3	4.0	1.3	1
5	6.3	2.5	4.9	1.5	1

Table 5.3. *Classification dataset*

5.3.3.1. *Construction of new datasets from original data*

The first step is to randomly generate new datasets from original data, represented in Tables 5.4–5.7.

For the sake of simplicity, we will create four datasets (over 100 are actually created). We will randomly select lines of the original dataset to make each new set. Each newly created dataset will involve the same number of lines as the original set.

It should be noted that it is possible that a line is generated several times in the same dataset, given that we do not make a random sampling with replacement.

id	x_0	x_1	x_2	x_3	y
2	5.4	3.9	1.7	0.4	0
0	4.4	2.9	1.4	0.2	0
2	5.4	3.9	1.7	0.4	0
4	5.5	2.3	4.0	1.3	1
5	6.3	2.5	4.9	1.5	1
5	6.3	2.5	4.9	1.5	1

Table 5.4. *Dataset 1*

id	x_0	x_1	x_2	x_3	y
2	5.4	3.9	1.7	0.4	0
1	5.3	3.7	1.5	0.2	0
3	6.6	3.0	4.4	1.4	1
1	5.3	3.7	1.5	0.2	0
4	5.5	2.3	4.0	1.3	1
4	5.5	2.3	4.0	1.3	1

Table 5.5. *Dataset 2*

id	x_0	x_1	x_2	x_3	y
4	5.5	2.3	4.0	1.3	1
1	5.3	3.7	1.5	0.2	0
3	6.6	3.0	4.4	1.4	1
0	4.4	2.9	1.4	0.2	0
0	4.4	2.9	1.4	0.2	0
2	5.4	3.9	1.7	0.4	0

Table 5.6. *Dataset 3*

id	x_0	x_1	x_2	x_3	*Y*
3	6.6	3.0	4.4	1.4	1
3	6.6	3.0	4.4	1.4	1
2	5.4	3.9	1.7	0.4	0
5	6.3	2.5	4.9	1.5	1
1	5.3	3.7	1.5	0.2	0
2	5.4	3.9	1.7	0.4	0

Table 5.7. *Dataset 4*

5.3.3.2. *Data training*

To train data, we will not use all of the available attributes. We will instead randomly select a specific subset of characteristics for each set and will use them for training.

To generate a decision tree for dataset 1, we will only use the attributes $\boldsymbol{x_0}$ and $\boldsymbol{x_1}$. Then, for dataset 2, we will use $\boldsymbol{x_2}$ and $\boldsymbol{x_3}$, for dataset 3, we will use $\boldsymbol{x_1}$ and $\boldsymbol{x_3}$, and finally for dataset 4, we will use $\boldsymbol{x_0}$ and $\boldsymbol{x_2}$.

Tables 5.8–5.11 represent the data subsets for datasets 1, 2, 3 and 4, respectively.

id	x_0	x_1	*y*
2	5.4	3.9	0
0	4.4	2.9	0
2	5.4	3.9	0
4	5.5	2.3	1
5	6.3	2.5	1
5	6.3	2.5	1

Table 5.8. *Data subset 1*

id	x_2	x_3	y
2	1.7	0.4	0
1	1.5	0.2	0
3	4.4	1.4	1
1	1.5	0.2	0
4	4.0	1.3	1
4	4.0	1.3	1

Table 5.9. *Data subset 2*

id	x_1	x_3	y
4	2.3	1.3	1
1	3.7	0.2	0
3	3.0	1.4	1
0	2.9	0.2	0
0	2.9	0.2	0
2	3.9	0.4	0

Table 5.10. *Data subset 3*

id	x_0	x_2	y
3	6.6	4.4	1
3	6.6	4.4	1
2	5.4	1.7	0
5	6.3	4.9	1
1	5.3	1.5	0
2	5.4	1.7	0

Table 5.11. *Data subset 4*

Figure 5.10 represents the decision trees for data subsets 1, 2, 3 and 4, as follows.

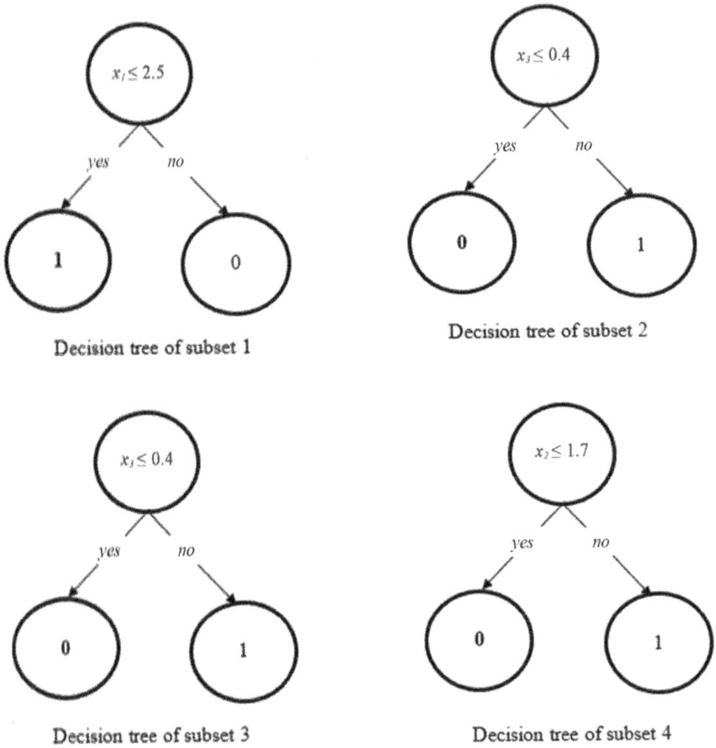

Figure 5.10. *Decision trees for data subsets 1, 2, 3 and 4. For a color version of this figure, see www.iste.co.uk/elhami/applicationsofai.zip*

5.3.3.3. *Making a prediction of a new datapoint*

Let us now consider a new datapoint, represented in Table 4.12. This point will be passed through each decision tree. For each tree, the generated prediction will be noted. Using a majority vote, we will predict the class of this point depending on the predictions of all trees.

x_0	x_1	x_2	x_3	y
2.3	1.0	0.6	0.1	?

Table 5.12. *New datapoint. For a color version of this table, see www.iste.co.uk/elhami/applicationsofai.zip*

Figure 5.11 illustrates the predictions provided by decision trees for each data subset.

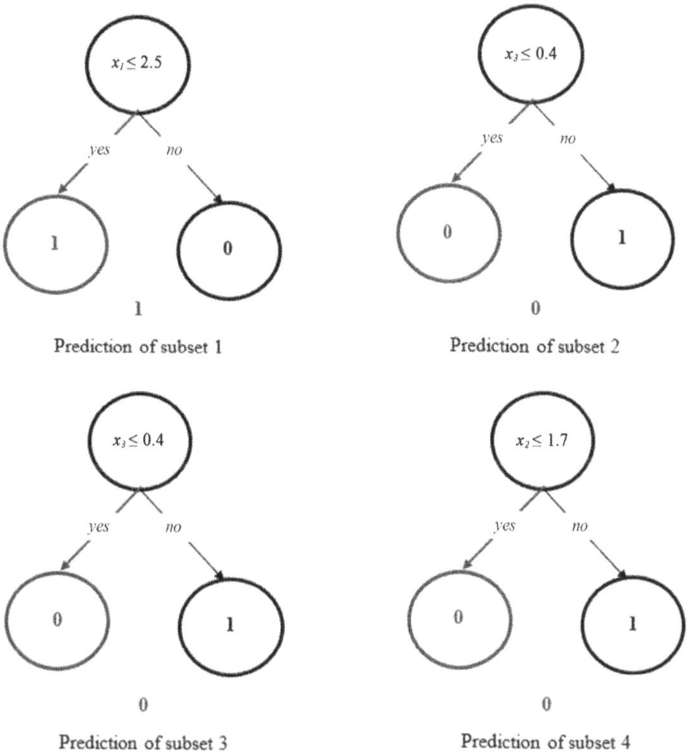

Figure 5.11. *Predictions of decision trees for data subsets. For a color version of this figure, see www.iste.co.uk/elhami/applicationsofai.zip*

After having combined all predictions, it can be noted that the majority vote yields the result 0. It can consequently be concluded that for this datapoint, the class y is 0, as shown in Table 4.13.

x_0	x_1	x_2	x_3	y
2.3	1.0	0.6	0.1	0

Table 5.13. *Prediction result. For a color version of this table, see www.iste.co.uk/elhami/applicationsofai.zip*

5.3.4. *Random forest example for regression*

In this example, similar to the previous one, we have a dataset containing the attributes x_0, x_1, x_2 and x_3 and also an explanatory variable y, as illustrated in Table 5.14. However, this time, the variable y is continuous and non-discrete. The stages of the algorithm will be applied similarly to the classification example. The only difference is in the calculation of the result predicted by the set of decision trees.

For the classification, we have obtained the prediction result thanks to the majority vote. This time, we will calculate it by determining the average of all predictions.

id	x_0	x_1	x_2	x_3	y
0	0.48	-1.18	-1.39	-1.31	-177.03
1	0.41	0.37	1.02	-1.39	-88.35
2	-0.15	0.19	1.64	0.82	118.63
3	-1.49	-1.08	0.36	1.11	62.42
4	0.13	-0.05	-2.30	0.53	9.94
5	2.10	-1.38	1.22	0.12	-14.46

Table 5.14. *Regression dataset*

5.3.4.1. *Data construction and training*

We will start by randomly selecting four new datasets, though in practice we often chose over 100 for our needs. Then, from these datasets, we will extract subsets. Tables 4.15–4.18 are assigned to new datasets, while Tables 4.19–4.22 represent the subsets of extracted data.

id	x_0	x_1	x_2	x_3	y
1	0.41	0.37	1.02	-1.39	-88.35
0	0.48	-1.18	-1.39	-1.31	-177.03
2	-0.15	0.19	1.64	0.82	118.63
0	0.48	-1.18	-1.39	-1.31	-177.03
4	0.13	-0.05	-2.30	0.53	9.94
5	2.10	-1.38	1.22	0.12	-14.46

Table 5.15. *Dataset 1*

id	x_0	x_1	x_2	x_3	y
5	2.10	-1.38	1.22	0.12	-14.46
4	0.13	-0.05	-2.30	0.53	9.94
2	-0.15	0.19	1.64	0.82	118.63
2	-0.15	0.19	1.64	0.82	118.63
4	0.13	-0.05	-2.30	0.53	9.94
5	2.10	-1.38	1.22	0.12	-14.46

Table 5.16. *Dataset 2*

id	x_0	x_1	x_2	x_3	y
5	2.10	-1.38	1.22	0.12	-14.46
4	0.13	-0.05	-2.30	0.53	9.94
3	-1.49	-1.08	0.36	1.11	62.42
3	-1.49	-1.08	0.36	1.11	62.42
4	0.13	-0.05	-2.30	0.53	9.94
5	2.10	-1.38	1.22	0.12	-14.46

Table 5.17. *Dataset 3*

id	x_0	x_1	x_2	x_3	y
4	0.13	-0.05	-2.30	0.53	9.94
4	0.13	-0.05	-2.30	0.53	9.94
2	-0.15	0.19	1.64	0.82	118.63
3	-1.49	-1.08	0.36	1.11	62.42
4	0.13	-0.05	-2.30	0.53	9.94
1	0.41	0.37	1.02	-1.39	-88.35

Table 5.18. *Dataset 4*

id	x_0	x_1	y
1	0.41	0.37	-88.35
0	0.48	-1.18	-17.03
2	-0.15	0.19	118.63
0	0.48	-1.18	-177.03
4	0.13	-0.05	9.94
5	2.10	-1.38	-14.46

Table 5.19. *Data subset 1*

id	x_1	x_2	y
5	-1.38	1.22	-14.46
4	-0.05	-2.30	9.94
2	0.19	1.64	118.63
2	0.19	1.64	118.63
4	-0.05	-2.30	9.94
5	-1.38	1.22	-14.46

Table 5.20. *Data subset 2*

id	x_2	x_3	y
5	1.22	0.12	-14.46
4	-2.30	0.53	9.94
3	0.36	1.11	62.42
3	0.36	1.11	62.42
4	-2.30	0.53	9.94
5	1.22	0.12	-14.46

Table 5.21. *Data subset 3*

id	x_0	x_3	y
4	0.13	0.53	9.94
4	0.13	0.53	9.94
2	-0.15	0.82	118.63
3	-1.49	1.11	62.42
4	0.13	0.53	9.94
1	0.41	-1.39	-88.35

Table 5.22. *Data subset 4*

Here are the decision trees corresponding to data subsets 1, 2, 3 and 4, illustrated in Figures 5.12–5.15.

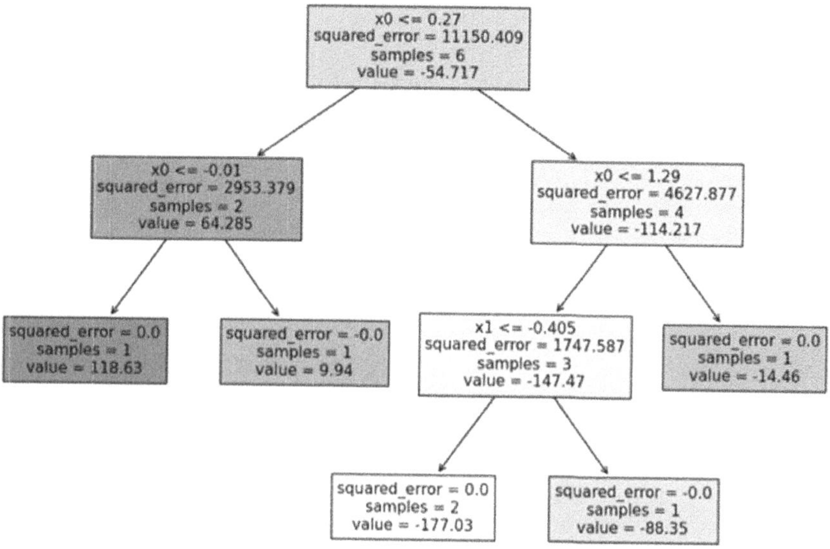

Figure 5.12. *Decision tree for data subset 1. For a color version of this figure, see www.iste.co.uk/elhami/applicationsofai.zip*

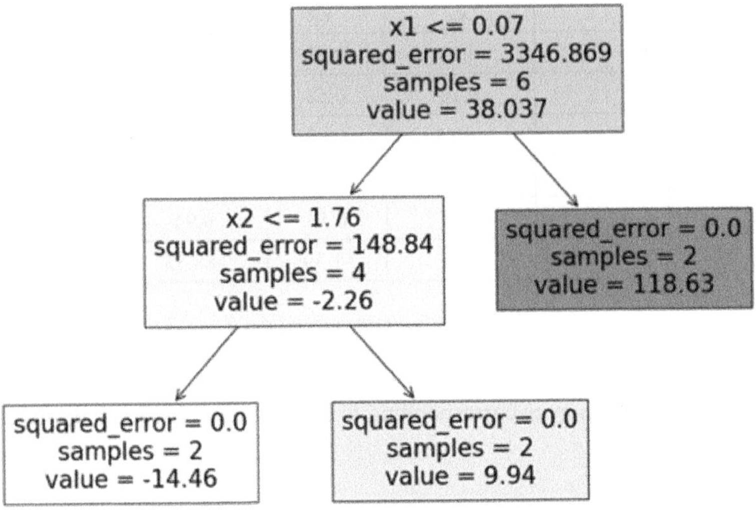

Figure 5.13. *Decision tree for data subset 2. For a color version of this figure, see www.iste.co.uk/elhami/applicationsofai.zip*

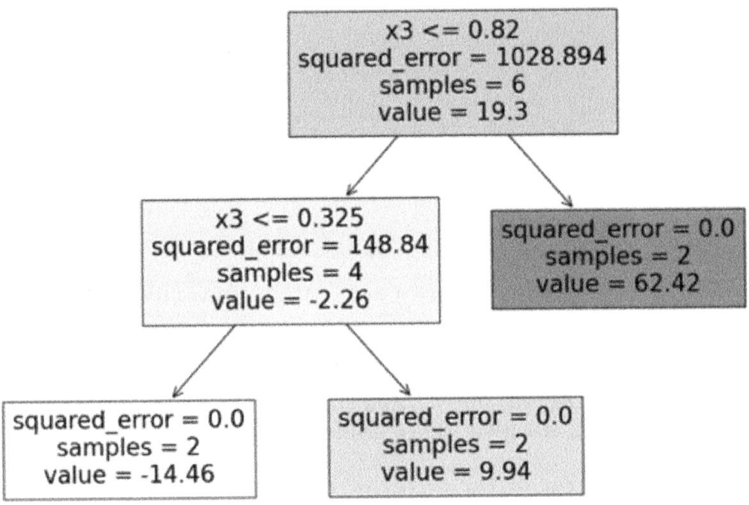

Figure 5.14. *Decision tree for data subset 3. For a color version of this figure, see www.iste.co.uk/elhami/applicationsofai.zip*

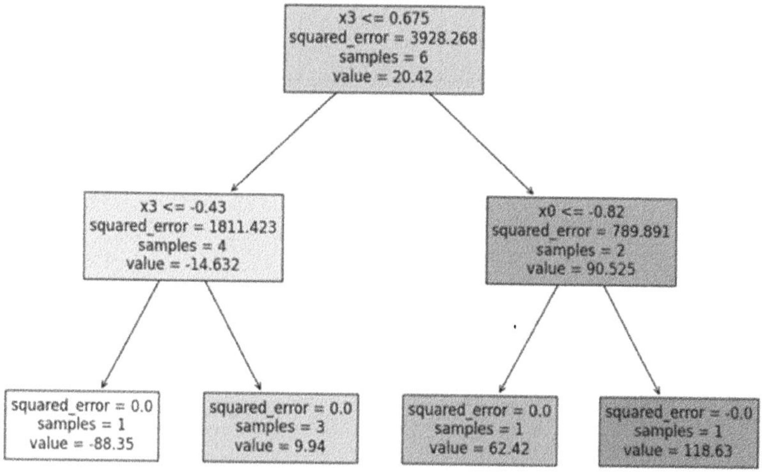

Figure 5.15. *Decision tree for data subset 4. For a color version of this figure, see www.iste.co.uk/elhami/applicationsofai.zip*

It should be noted that these decision trees are constructed by minimizing the quadratic error. This construction process generally includes several key stages. Initially, the initial dataset is divided into subsets, often in a recursive manner. At each stage of this division, the algorithm aims to select the division variable that minimizes the quadratic error. The latter is calculated by assessing the gap between the values predicted by the decision tree and the actual values in each subset. For this purpose, the algorithm explores various division variables and also various cut-off points for each of them in order to determine the division that minimizes the error. Once an optimal division is identified, the process is repeated for each subset created by the previous division, until a stop criterion is reached, such as a minimal size of the subset or a maximal number of tree levels. This iterative approach allows the decision tree to fit the data and grasp the complex relations between the variables, while minimizing the global quadratic error of the regression model (Jain and Jana 2023).

5.3.4.2. *Predicting a new datapoint*

Let us consider a new datapoint to be predicted, represented in Table 5.24. This point will be passed through each decision tree. The prediction generated by each tree is recorded. This time, for the prediction, we determine the class of this point by taking the average of predictions of all trees. The results of these predictions are represented in Figures 5.16–5.19.

x_0	x_1	x_2	x_3	Y
-0.07	-0.30	-0.12	0.10	?

Table 5.23. *New datapoint. For a color version of this figure, see www.iste.co.uk/elhami/applicationsofai.zip*

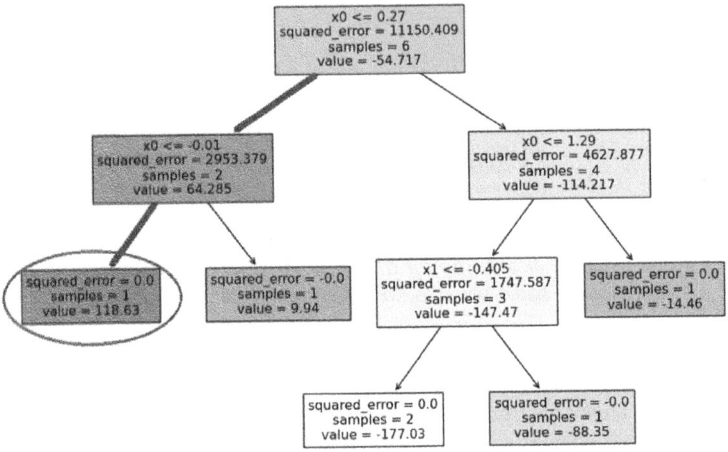

Figure 5.16. *Prediction of results for data subset 1. For a color version of this figure, see www.iste.co.uk/elhami/applicationsofai.zip*

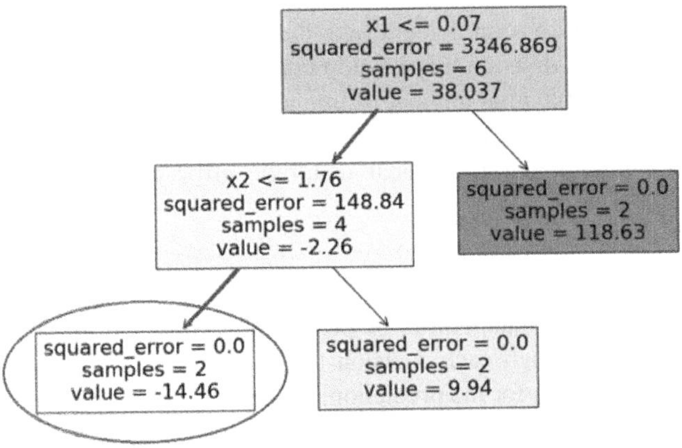

Figure 5.17. *Prediction of results for data subset 2. For a color version of this figure, see www.iste.co.uk/elhami/applicationsofai.zip*

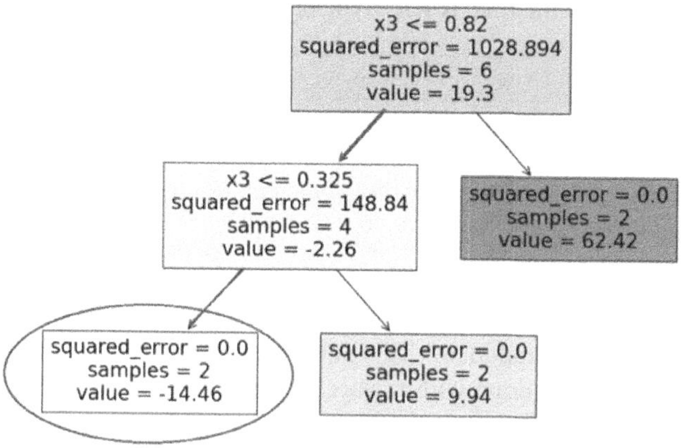

Figure 5.18. *Prediction of results for data subset 3. For a color version of this figure, see www.iste.co.uk/elhami/applicationsofai.zip*

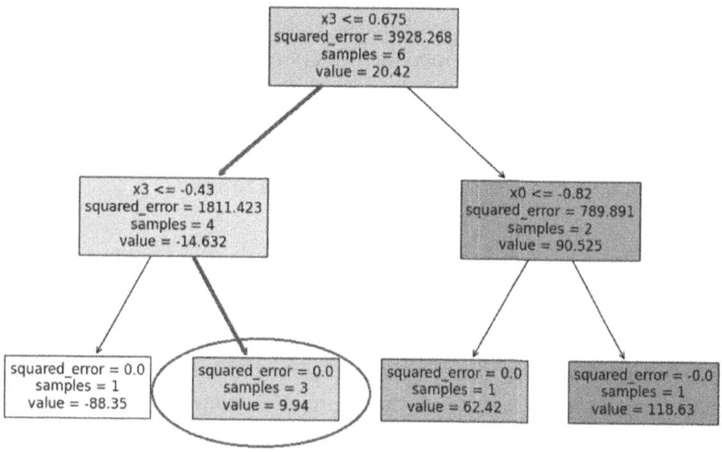

Figure 5.19. *Prediction of results for data subset 4. For a color version of this figure, see www.iste.co.uk/elhami/applicationsofai.zip*

The results of predictions are: (188.63), (-14.46), (-14.46) and (9.94)

The prediction of y is therefore: $\frac{188.63 - 14.46 - 14.46 + 9.94}{4}$; $y = \mathbf{42.412}$

Table 5.24 presents the final result of the prediction.

x_0	x_1	x_2	x_3	y
-0.07	-0.30	-0.12	0.10	42.412

Table 5.24. *Final result of the prediction. For a color version of this figure, see www.iste.co.uk/elhami/applicationsofai.zip*

NOTE.– Some crucial aspects of the random forest algorithm should be addressed. Firstly, why is it called "random"? This is because we used two random processes: bootstrapping (randomly draw with replacement of samples from an initial training dataset) and random selection of characteristics. But what is the reason for bootstrapping and random selection of characteristics? Bootstrapping ensures the use of the same data for each tree, which renders our model less sensitive to the original training data. The random selection of characteristics helps us to reduce the correlation between trees. If all of the characteristics are used, most trees will have similar decision nodes, thus increasing the variance. This represents another advantage of the random selection of characteristics. Some trees will be based on less important characteristics and will hence yield poor predictions, but other trees will yield poor predictions in an opposite direction, which will balance predictions. Then, what is the ideal size of the subset of characteristics? In our case, we chose two characteristics corresponding to the square root of the total number of characteristics. Research showed that the values near the logarithm or the square root of the total number of characteristics yield good results.

5.4. Case study: detection of malware on Android

Given the increasing threat of malware targeting mobile devices, it is a necessity to design efficient methods for detecting and disabling these harmful programs. From this perspective, we present a database containing the characteristics of several mobile applications. We will use the random forest algorithm on this database to determine if an application has a malicious nature or not. In order to assess the performance of our algorithm, we will divide our database into two sets: 80% of data will be dedicated to algorithm training, while the remaining 20% will serve to tests. We will then compare the results obtained by the algorithm with the test data to assess its accuracy.

5.4.1. *Data presentation*

The database consists of characteristic vectors including 215 attributes extracted from 15,036 applications (5,560 malicious ones issued from the Drebin project and

9,476 benign). It was used to develop and assess a multilevel classifier fusion approach aimed at detecting malware on Android. This research was published in the article entitled: "*DroidFusion: A novel multilevel classifier fusion approach for Android malware detection*" in IEEE Transactions on cybernetics (Yerima and Sexer 2019). Table 4.25 contains a description of characteristic vectors resulting from the analysis of the static code of Android applications.

Characteristic	Description
API call signature	An API call signature refers to the single identifier or the model representing a specific method or a function call to an external service or resource externe by means of an application programming interface (API).
Manifest permission	A manifest permission in Android is a declaration in an app's configuration file ("AndroidManifest.xml") that specifies the system features or data the app needs access to, such as the camera, Internet or contacts.
Intent	An intent is a mail object to require an action to another component of the application, generally between various activities, services, diffusion receivers or the system. Intents facilitate communication between the components of an Android application or between various applications.
Command signature	A command signature generally refers to the structure or format of a command used to conduct a specific action or operation in an Android application. These may include various functionalities such as data transmission between various components of the application, execution of specific tasks or interaction with the material or the functions of the device system.

Table 5.25. *Characteristics of attributes*

All of the attributes have two possible values, either 1 or 0 (Boolean data type). The class also takes two values: **B for "benign"**, which indicates that the application is friendly and **S for "malware"**, meaning that the application is a potential threat.

5.4.2. *Results*

After a preliminary clean-up of our data by removing inputs containing missing values, which reduced the total number of inputs to 15,031, we proceeded to the application of our algorithm to a set of data representing 80% of the total, which is

12,024 inputs. To assess our model performance, the remaining 20% were used as a set of tests where we made our predictions. Consequently, we calculated several metrics to assess the quality of our predictions, namely accuracy, sensitivity and the F-score.

Accuracy measures the proportion of instances predicted as positive that are actually positive. In other terms, it indicates the capacity of the model to correctly identify the true positive instances among all instances predicted as positive.

As for sensitivity, it measures the proportion of actually positive instances that were correctly predicted by the model. It highlights the capacity of the model to identify all of the true positives among all of the actually positive instances of the dataset.

The F-score is a measure that combines both accuracy and sensitivity into a single value. It represents the mean harmonic of accuracy and sensitivity. Hence, the F-score offers a balanced view between the capacity of the model to equally minimize false positives and detect all true positives.

The application of the algorithm to the random forest provided remarkable results, as proven by the achieved performances. The accuracy of predictions varies between an impressive minimum of 98.17% and a maximum of 99.33%, with a global average of 98.81% over 100 iterations, as illustrated in Figure 5.20. Similarly, the sensitivity of results proved to be robust, displaying a minimum of 97.71% and a maximum of 99.20%, with an average of 98.54%, as detailed in Figure 5.21. Finally, the F-score presents a similar coherence, with a minimum of 97.98%, a maximum of 99.28% and an average of 98.54%, as represented in Figure 5.22. These high performances prove the effectiveness of the algorithm in solving the problem under study.

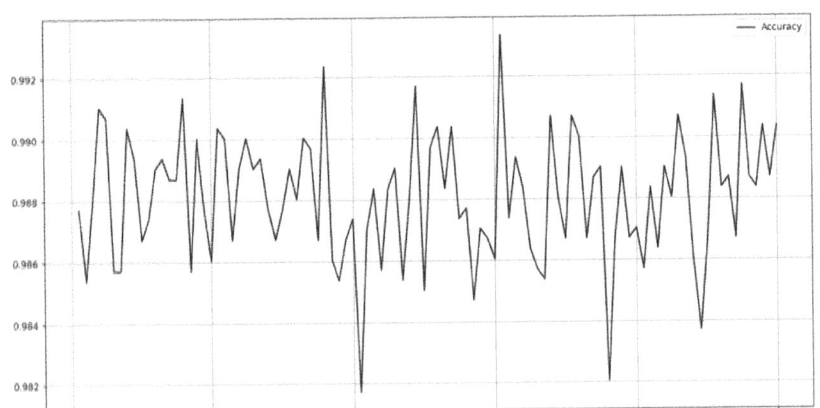

Figure 5.20. *Accuracy of the algorithm over 100 iterations*

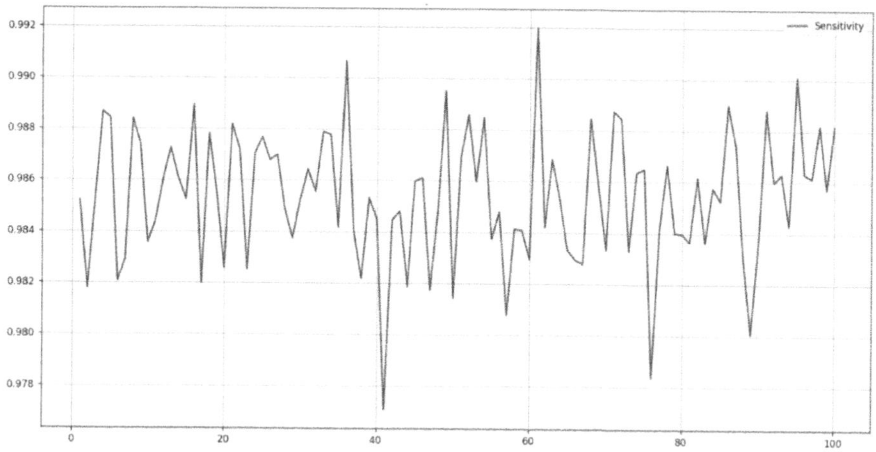

Figure 5.21. *Sensitivity of the algorithm over 100 iterations*

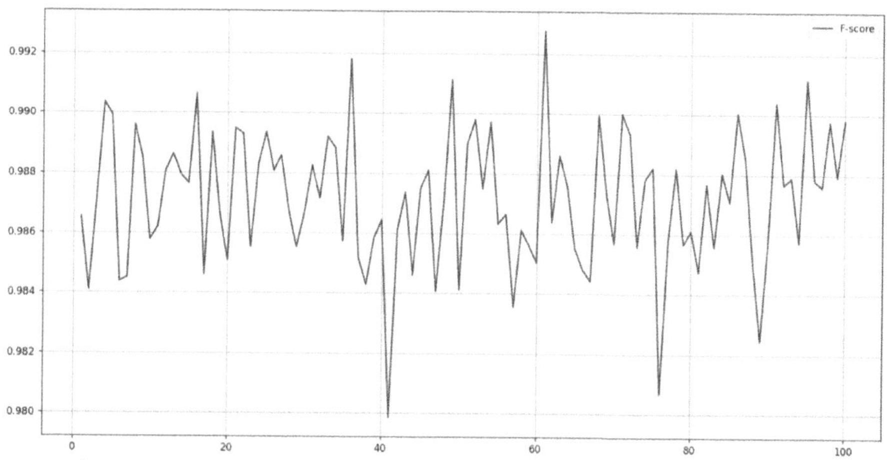

Figure 5.22. *F-score of the algorithm over 100 iterations*

5.5. Chapter essentials

This chapter presents a thorough exploration of the random forest algorithm, a powerful and multipurpose machine learning technique. We started with a general introduction that helped us to set the scene of this method and present its challenges. Then, we dived into the world of decision trees, the foundation of the random forest. This section explored how decision trees are the building blocks of this algorithm, thus providing a fundamental comprehension of its operation.

We continued our exploration by approaching the foundations of the random forest, detailing the underlying mechanisms through which this algorithm combines several decision trees to achieve robust and accurate predictions. We highlighted the advantages and the distinctive characteristics of this approach compared to the standard methods of machine learning.

Moreover, examples of practical applications for the decision trees and the random forest contributed to our understanding of these concepts and their utility in various domains.

Finally, an exciting case study was used as concrete illustration of the efficiency of random forest: the detection of malware on Android. This study proved how the principles of random forest can be successfully applied to solve actual problems, in this case, the detection of threats in the ecosystem of Android applications.

This chapter offered a complete and in-depth overview of the random forest algorithm and its components, essential to concrete applications resulting, thus illustrating its importance in the field of machine learning and beyond.

5.6. References

Abo Al-Kheer, A., Eid, M., Aoues, Y., El Hami, A., Kharmanda, M.G., Mouazen, A.M. (2011). Theoretical analysis of the spatial variability in tillage forces for fatigue analysis of tillage machines. *Journal of Terramechanics*, 48(4), 285–295.

Anas, E.A. and El Hami, N. (2022). Assignment problem solved by two metaheuristic algorithms ACO and HHO. In *2022 8th International Conference on Optimization and Applications (ICOA)*, Genoa. doi: 10.1109/ICOA55659.2022.9934141.

Antoniadis, A., Lambert-Lacroix S., Poggi, J. (2021). Random forests for global sensitivity analysis: A selective review. *Reliability Engineering & System Safety*, 206, 107312. doi: 10.1016/j.ress.2020.107312.

Belagoune, S., Bali, N., Bakdi, A., Baadji, B., Atif, K. (2021). Deep learning through LSTM classification and regression for transmission line fault detection, diagnosis and location in large-scale multi-machine power systems. *Measurement*, 177, 109330. doi: 10.1016/j.measurement.2021.109330.

Ben Abdessalem, A. and El Hami A. (2015). A probabilistic approach for optimising hydroformed structures using local surrogate models to control failures. *International Journal of Mechanical Sciences*, 96–97, 143–162.

Ben Abdessalem, A., Pagnacco, E., El Hami, A. (2013). Increasing the stability of T-shape tube hydroforming process under stochastic framework. *International Journal of Advanced Manufacturing Technology*, 69(5–8), 1343–1357.

Bouchekourte, M. and El Hami, N. (2022). Optimization of equity allocations of institutional investors: Study of Moroccan case. *International Journal for Simulation and Multidisciplinary Design Optimization*, 13, 12. doi: 10.1051/smdo/2021042.

Bouyekhf, R., El Moudni, A., El Hami, A., Zerhouni, N., Ferney, M. (1996). Reduced order modelling of two-time-scale discrete non-linear systems. *Journal of the Franklin Institute*, 333(4), 499–512.

Bouyekhf, R., El Hami, A., El Moudni, A. (2001). Optimal control of a particular class of singularly perturbed nonlinear discrete-time systems. *IEEE Transactions on Automatic Control*, 46(7), 1097–1101.

Castillo-Botón, C., Casillas-Pérez, D., Casanova-Mateo, C., Ghimire, S., Cerro-Prada, E., Gutierrez, P.A., Deo, R.C., Salcedo-Sanz, S. (2022). Machine learning regression and classification methods for fog events prediction. *Atmospheric Research*, 272, 106157. doi: 10.1016/j.atmosres.2022.106157.

Chanmee, S. and Kesorn, K. (2023). Semantic decision Trees: A new learning system for the ID3-Based algorithm using a knowledge base. *Advanced Engineering Informatics*, 58, 102156. doi: 10.1016/j.aei.2023.102156.

El Attaoui, A., Boukhalfa, A., Rhouas, S., El Hami, N. (2023). Custom application programming interface data extractor applied to the Klarna e-commerce dataset. *Indonesian Journal of Electrical Engineering and Computer Science*, 31(3). doi: 10.11591/ijeecs.v31.i3.pp1624–1632.

El Hami, A. and Radi, B. (1996). Some decomposition methods in the analysis of repetitive structures. *Computers and Structures*, 58(5), 973–980.

El Hami, A., Lallement, G., Minotti, P., Cogan, S. (1993). Methods that combine finite group theory with component mode synthesis in the analysis of repetitive structures. *Computers and Structures*, 48(6), 975–982.

El Hami, N., Ellaia, R., Itmi, M. (2010). Hybrid evolutionary optimization algorithm MPSO-SA. *International Journal for Simulation and Multidisciplinary Design Optimization*, 4(1), 27–32. doi: 10.1051/ijsmdo/2010004.

El Hami, N., Ellaia, R., Itmi, M. (2011a). A new methodology for the reliability based particle swarm optimization with simulated annealing. *Advanced Materials Research*, 274, 101–111. doi: 10.4028/www.scientific.net/amr.274.101.

El Hami, N., Itmi, M., Ellaia, R. (2011b). Reliability-based design and heuristic optimization MPSO-SA of structures. *Advanced Materials Research*, 274, 91–100. doi: 10.4028/www.scientific.net/amr.274.91.

El Hami, N., Elmir, A., Elmir, B., Bounabat, B. (2013). Simulation and optimization of interoperability planning. *Journal of Theoretical and Applied Information Technology*, 52(2), 348–356.

El Hami, N., Itmi, M., El Hami, A. (2015a). Hybrid evolutionary optimization algorithm for structures. *Advanced Materials Research*, 1099, 102–109. doi: 10.4028/www.scientific.net/amr.1099.102.

El Hami, N., Itmi, M., El Hami, A. (2015b). Simulation and optimization of an actuator. *Advanced Materials Research*, 1099, 110–119. doi: 10.4028/www.scientific.net/amr.1099.110.

El Hami, N., Ammari, N., Mrabti, A.A.E., Kalam, A.A.E., Ouahman, A.A. (2017a). Sécurisation de l'environnement mobile Combinaison du Framework MPSS et le Firewall ALSD. *Incertitudes et fiabilité des systèmes multiphysiques*, 17(1). doi: 10.21494/iste.op.2017.0128.

El Hami, N., Hmina, N., Koulou, A., Elmir, A. (2017b). Une nouvelle méthode pour l'évaluation de l'interopérabilité des systems. *Incertitudes et fiabilité des systèmes multiphysiques*, 17(1). doi: 10.21494/iste.op.2017.0125.

Hakima, R., Maria, Z., Norelislam, E.H., Nabil, H. (2022). A comparative study of several metaheuristic algorithms for optimization problems. In *2022 8th International Conference on Optimization and Applications (ICOA)*, Genoa. doi: 10.1109/ICOA55659.2022.9934204.

Hasan, J. and Horvat, M. (2024). An application of the Random Forest algorithm for the prediction of Solar Envelope 'Floor Space Index' based on spatiotemporal parameters. *Journal of Building Engineering*, 86, 108784. doi: 10.1016/j.jobe.2024.108784.

Jain, N. and Jana, P.K. (2023). LRF: A logically randomized forest algorithm for classification and regression problems. *Expert Systems with Applications*, 213, 119225. doi: 10.1016/j.eswa.2022.119225.

Kharmanda, G., Ibrahim, M.-H., Abo Al-kheer, A., Guerin, F., El Hami, A. (2014). Reliability-based design optimization of shank chisel plough using optimum safety factor strategy. *Computers and Electronics in Agriculture*, 109, 162–171.

Koulou, A., Zemzami, M., El Hami, N., Elmir, A., Hmina, N. (2020). Optimization in collaborative information systems for an enhanced interoperability network. *International Journal for Simulation and Multidisciplinary Design Optimization*, 11, 2. doi: 10.1051/smdo/2019021.

Li, P., Xiong, F., Huang, X., Wen, X. (2024). Construction and optimization of vending machine decision support system based on improved C4.5 decision tree. *Heliyon*, 10(3), e25024. doi: 10.1016/j.heliyon.2024.e25024.

Mabrouk, I.B. and El Hami, A. (2019). Effect of number of blades on the dynamic behavior of a Darrieus turbine geared transmission system. *Mechanical Systems and Signal Processing*, 121, 562–578.

Makhloufi, A., Aoues, Y., El Hami, A. (2016). Reliability based design optimization of wire bonding in power microelectronic devices. *Microsystem Technologies*, 22(12), 2737–2748.

Mojsilović, M., Cvejić, R., Pepić, S., Karabašević, D., Saračević, M., Stanujkić, D. (2023). Statistical evaluation of the achievements of professional students by combination of the random forest algorithm and the ANFIS method. *Heliyon*, 9(11), e21768. doi: 10.1016/j.heliyon.2023.e21768.

Moro, T., El Hami, A., El Moudni, A. (2002). Reliability analysis of a mechanical contact between deformable solids. *Probabilistic Engineering Mechanics*, 17(3), 227–232.

Najim, A., Ghallali, M., Kalam, A.A., El Hami, N., Ouahman, A.A., Ouahidi, B.E. (2014). Mobile security: Security mechanisms and protection of mobile applications. *Journal of Theoretical & Applied Information Technology*, 70, 302–315.

Rhouas, S. and El Hami, N. (2022). A packaging industry optimization based on three metaheuristics methods. In *2022 8th International Conference on Optimization and Applications (ICOA)*, Genoa. doi: 10.1109/ICOA55659.2022.9934635.

Rhouas, S., Bouchekourte, M., El Hami, A. (2022). Optimization of the impact measurement of market structure on liquidity and volatility. *International Journal for Simulation and Multidisciplinary Design Optimization*, 13(9). doi: 10.1051/smdo/2021040.

Toft, H.B., Müller, K., Hendrikx, J., Jaedicke, C., Bühler, Y. (2023). Can big data and random forests improve avalanche runout estimation compared to simple linear regression? *Cold Regions Science and Technology*, 211, 103844. doi: 10.1016/j.coldregions.2023.103844.

Wan, M., Wu, Q., Yan, L., Guo, J., Li, W., Lin, W., Lu, S. (2023). Taxi drivers' traffic violations detection using random forest algorithm: A case study in China. *Traffic Injury Prevention*, 24(4), 362–370. doi: 10.1080/15389588.2023.2191286.

Wang, C., Wang, Y., Xu, J., Wang, Q., Zhou, F. (2020). Outsourced privacy-preserving decision tree classification service over encrypted data. *Journal of Information Security and Applications*, 53, 102517. doi: 10.1016/j.jisa.2020.102517.

Yan, T., Rui, X., Shi-Hui, S., Zhao-Kai, H., Jin-Yu, F. (2023). A real-time intelligent lithology identification method based on a dynamic felling strategy weighted random forest algorithm. *Petroleum Science.* doi: 10.1016/j.petsci.2023.09.011.

Yerima, S.Y. and Sexer, S. (2019). DroidFusion: A novel multilevel classifier fusion approach for android malware detection. *IEEE Transactions on Cybernetics*, 49(2), 453–466. doi: 10.1109/tcyb.2017.2777960.

Zeine, A.T., El Hami, N., El Hami, A., Ouhimmou, S., Ellaia, R. (2017). Multiobjective optimization of trusses using Backtracking Search Algorithm. *Uncertainties and Reliability of Multiphysical Systems*, 17(1), doi: 10.21494/iste.op.2017.0121.

Zemzami, M. (2019). A modified particle swarm optimization algorithm linking dynamic neighborhood topology to parallel computation. *International Journal of Advanced Trends in Computer Science and Engineering*, 8(2), 112–118. doi: 10.30534/ijatcse/2019/03822019.

Zemzami, M., El Hami, N., Makhloufi, A., Itmi, M., Hmina, N. (2017). Application d'un modèle parallèle de la méthode PSO au problème de transport d'électricité. *Incertitudes et fiabilité des systèmes multiphysiques*, 17(1). doi: 10.21494/iste.op.2017.0127.

Zemzami, M., El Hami, N., Itmi, M., Hmina, N., Lgs-Ensa-Kenitra, M. (2019). An evolutionary hybrid algorithm for complex optimization problems. *International Journal of Advanced Trends in Computer Science and Engineering*, 8(2), 126–133. doi: 10.30534/ijatcse/2019/05822019.

Zemzami, M., El Hami, N., Itmi, M., Hmina, N. (2020). A comparative study of three new parallel models based on the PSO algorithm. *International Journal for Simulation and Multidisciplinary Design Optimization*, 11, 5. doi: 10.1051/smdo/2019022.

Zhang, X., Shen, H., Huang, T., Wu, Y., Guo, B., Liu, Z., Luo, H., Tang, J., Zhou, H., Wang, L., Xu, W., Ou, G. (2024). Improved random forest algorithms for increasing the accuracy of forest aboveground biomass estimation using Sentinel-2 imagery. *Ecological Indicators*, 159, 111752. doi: 10.1016/j.ecolind.2024.111752.

Zhao, F., Tang, Z., Xiao, Z., Liu, H., Fan, J., Li, L. (2024). Ensemble CART surrogate-assisted automatic multi-objective rough fuzzy clustering algorithm for unsupervised image segmentation. *Engineering Applications of Artificial Intelligence*, 133, 108104. doi: 10.1016/j.engappai.2024.108104.

Zhou, J., Yang, P., Li, C., Du, K. (2023). Hybrid random forest-based models for predicting shear strength of structural surfaces based on surface morphology parameters and metaheuristic algorithms. *Construction and Building Materials*, 409, 133911. doi: 10.1016/j.conbuildmat.2023.133911.

6

Linear Regression and Application to Artificial Intelligence

6.1. Introduction

The current age is characterized by an unparalleled technological revolution centered on artificial intelligence (AI), which plays an essential role. With the radical transformation of a wide variety of sectors such as health, finance and engineering, AI pushes the limits of innovation and human capability. Its strength lies in its outstanding ability to analyze complex data, to use this data to build models and predict results with an accuracy that is often beyond human comprehension. This revolutionary capacity to transform raw data into accurate and operable knowledge marks the beginning of a new age in research and technological application (Lee et al. 2023).

Linear regression is a fundamental pillar in the foundation of this advanced AI universe. Though it may seem simplistic compared to other latest generation machine learning models, its value should not be underestimated. Linear regression offers a clear and direct method to identify the relations between variables, thus providing a solid basis for data interpretation and prediction. It is this initial context of being easy to understand and apply that allows data scientists and researchers to make reliable predictions and construct more complex models (Poix and Solev 2001).

The interaction between AI and linear regression is the materialization of a perfect synergy between simplicity and complexity. While the AI strives to navigate the ocean of data and their often nonlinear interactions, linear regression serves as an anchor, being used for the initial exploration and understanding of tendencies.

Chapter written by Sara RHOUAS, Norelislam EL HAMI and Younes KOULOU.

This symbiotic relation enhances the arsenal of AI, and advanced techniques such as polynomial regression, Ridge regression and LASSO regression can be used to overcome the limitations of simple linear regression, while preserving its fundamental philosophy of clear data interpretation (El Attaoui et al. 2023).

This chapter presents a detailed exploration of linear regression, its operation and practical application in the context of AI. We will show how this fundamental statistical method can be used to analyze the relations between a dependent variable and one or more independent variables, which is crucial for modeling AI data. We will also examine the place of linear regression in the wider context of regression analysis, aiming to model and understand these relations in the context of supervised learning algorithms. In particular, we will see how linear regression tries to establish a linear relation between variables, assuming that the changes in the dependent variable are proportional to the changes in independent variables. This will be illustrated with a practical example to facilitate the understanding of its concrete application to various AI domains, such as time series forecast, data analysis and predictive modeling.

6.2. Linear regression model

The linear regression model is an essential tool in statistical analysis for the study of relations between variables. By decomposing data into explanatory components, this model can be used to understand and predict the behavior of a dependent variable as a function of independent variables. With the rise of AI, these models can be enhanced and automated for more accurate analysis and more efficient prediction.

6.2.1. *Objectives and use*

The objective of linear regression, a powerful statistical method, is mainly to explore and quantify relations between variables. Its first objective is to understand the relation between a dependent variable and one or more independent variables, which serves to predict the values of the dependent variable as a function of the values of independent variables. Due to this capacity, it is an essential tool in many fields of research and data analysis. For example, in the public health domain, linear regression can be used to examine the factors that influence the prevalence of a disease, such as lifestyle, environment or access to health care (Demirci et al. 2023).

Another major objective of linear regression is to make predictions and assessments. The relations identified by linear regression can be used to predict the future values of the dependent variable as a function of the current or projected values of independent variables. This capacity is particularly valuable in fields such as finance, in which economic trend forecasts are crucial for informed decision-making (Sangeetha and Alfia 2023).

Furthermore, linear regression is broadly used in academic research for hypotheses testing and theory assessment. By building regression models, researchers can analyze the statistical importance of the relations between variables and assess the validity of their hypotheses. This facilitates the in-depth comprehension of the studied phenomena and contributes to the progress of knowledge in various disciplinary fields. In conclusion, given its versatility and capacity to model complex relations, linear regression continues to be widely used to address a broad range of research subjects and analytical problems (Liu et al. 2022).

6.2.2. *Applications to everyday life*

Linear regression is a fundamental tool widely used in various fields, from economy to medicine and passing through education. Its main objective is to model and understand the relations between variables, which facilitates addressing complex matters of everyday life. In the field of economy, it is crucial to analyze economic trends and assess the impact of public policies on economic stability and social well-being. For example, it can be used for studying the relation between unemployment rate and economic growth, thus providing essential insights for decision-makers (Vanhoucke and Colin 2016).

In medicine, linear regression is a valuable tool for studying the connections between risk factors and clinical results. For example, it may help in assessing the pollution exposure effect on respiratory health or in predicting the risk to developing certain diseases depending on family history and lifestyle. These analyses are crucial for elaborating efficient prevention and treatment strategies, thus contributing to public health improvement (Jahandideh et al. 2009).

In the field of education, linear regression can be used to analyze the factors that influence students' academic performances. For example, it can be used to assess the impact of teaching resources on standardized test results or to predict examination scores as a function of study time and student motivation level. This information helps educators target the interventions required to improve student results, thus contributing to strengthening education systems. In summary, due to its capacity to model complex relations between variables, linear regression plays an essential role in solving practical problems and in knowledge advancement in various fields of everyday life (Pavelescu 2015).

6.2.3. *Simple model vs. multiple model*

In the context of linear regression, the comprehension of dependent and independent variables, as well as their mathematical relation is crucial for model construction. Dependent variables, generally denoted by Y, are often at the core of the linear regression model, as they are the variables to be explained or predicted. For example, in a salary study, the dependent variable could be an individual's monthly salary. On the other hand, the independent variables are those assumed to influence the dependent variable. In this example, the independent variables could be the level of education, the professional experience and the field of employment (Wang et al. 2024).

This distinction between dependent and independent variables leads to the differentiation between two types of linear regression models: the simple model and the multiple model. The simple model is used when there is a single independent variable for predicting the dependent variable. Mathematically, it postulates a linear relation between the dependent variable and the independent variable (Ren 2022). The simple linear regression equation can be represented as follows:

$$Y = \beta_0 + \beta_1 X_1 + \varepsilon \qquad [6.1]$$

where Y is the dependent variable, X_1 is the independent variable, β_0 is the ordinate at the origin, β_1 is the regression coefficient associated with X_1 and ε is the random error.

This model is often used to study the relation between two variables and is relatively simple to interpret. On the other hand, the multiple model is used when several independent variables are used to predict the dependent variable (Olsen et al. 2020). In this case, the multiple linear regression equation is more complex, including several independent variables and their corresponding regression coefficients, as follows:

$$Y = \beta_0 + \beta_1 X_1 + \beta_2 X_2 + ... + + \beta_n X_n + \varepsilon \qquad [6.2]$$

where X_i represents an independent variable and β_i is the corresponding regression coefficient. By using this approach, the simultaneous effect of several variables on the dependent variable can be taken into account, leading to more accurate predictions. However, the interpretation of coefficients in the multiple model is more complex than in the simple model, as each coefficient measures the effect of the corresponding variable on the dependent variable while taking into account other independent variables in the model (Zhang et al. 2024).

In conclusion, the choice between the simple model and the multiple model depends on data complexity and on the specific research question. Each model has unique advantages and drawbacks, and the researcher must select the one that best fits its analytical objectives and the structure of available data (Hanley 2016).

6.2.4. *Regression coefficients*

Regression coefficients play a crucial role in the analysis of linear regression, as they are estimated parameters that capture the force and direction of the relation between the dependent variable and each independent variable of the model. In a simple model, a single regression coefficient is determined, representing the slope of the regression line. This coefficient reflects how the dependent variable changes on average for a unit change in the independent variable, while maintaining the other variables constant (Liu et al. 2020).

However, in a multiple model, each independent variable has its own regression coefficient. These coefficients measure the effect of each variable on the dependent variable while taking into account the influence of other variables of the model. Hence, these coefficients can be used to quantify the relative importance of each variable in explaining variations of the dependent variable, thus offering a more in-depth comprehension of the complex relations between various variables (Huang and Hsieh 2020).

The linear regression model is a powerful tool for exploring and analyzing the relations between variables. By correct identification of dependent and independent variables, selection of the appropriate type of model (simple or multiple), and interpretation of the regression coefficients, researchers are able to better understand and predict the behavior of the studied phenomena, thus opening a path to more detailed analyses and more accurate predictions (Prion and Adamson 2020).

In parallel with standard analyses, optimization and AI have revolutionized our manner of approaching linear regression. Indeed, these approaches can be used to automate and significantly improve the modeling process (Koulou et al. 2020; Bouchekourte et al. 2022). Optimization techniques such as gradient descent can be used to efficiently adjust the regression coefficients to minimize the prediction error, thus leading to more accurate and more reliable models (Bouchekourte and El Hami 2022; Rhouas et al. 2022).

Furthermore, the integration of AI, particularly through machine learning algorithms, can be used to identify complex and nonlinear models in the data, going beyond the limits of standard linear models (Rhouas et al. 2023).

By combining the fundamental principles of linear regression with the capacities of optimization and AI, researchers are now able to explore and model complex systems with unprecedented accuracy and depth (El Hami et al. 2010; Rhouas et al. 2022). This integrated approach opens new perspectives for the comprehension and prediction of studied phenomena, thus offering powerful tools for decision-making and problem solving in various domains.

6.3. Methods for assessing coefficients

As a fundamental statistical method for modeling relations between variables, linear regression requires effective techniques for assessing the coefficients of the model. This section focuses on commonly used methods.

6.3.1. *Ordinary least squares (OLS) method*

The ordinary least squares (OLS) method is the most commonly used method for assessing regression coefficients in a linear model. It is widely used due to its simplicity and ease of interpretation. Optimization plays a crucial role in this method, as it consists of adjusting the regression coefficients to find the best possible fitting of the model to data (Zeine et al. 2017; Zemzami et al. 2018). It is based on the minimization of the sum of squared residuals (SSR), also known as sum of squared errors (SSE):

$$SSE = \sum_{i=1}^{n} \left(Y_i - \widehat{Y}_i \right)^2 \qquad [6.3]$$

The sum of squared errors is defined as the sum of squared gaps between the observed values and the values predicted by the model, thus providing a measure of the overall fitting of the model to data. Mathematically, the OLS method consists of solving the following optimization problem:

$$\min_{\beta} \sum_{i=1}^{n} \left(y_i - \beta_0 - \beta_1 x_{i1} - \ldots - \beta_p x_{ip} \right)^2 \qquad [6.4]$$

where y_i represents the observed values, β is the vector of coefficients to be assessed and x_{ij} are the predictors. The coefficients can be assessed by solving the normal equations:

$$\widehat{\beta} = \left(X^T X \right)^{-1} X^T y \qquad [6.5]$$

where X is the design matrix, with each line representing an observation and each column an explanatory variable, and y is the vector of observed values.

This mathematical formulation can be used to obtain the optimal coefficients that minimize the sum of squared residuals, thus providing the best assessments of model parameters. The main advantage of OLS is its design simplicity and ease of implementation, which makes it an accessible tool even for beginners in statistics. Furthermore, the estimators of coefficients obtained by OLS have well-established statistical properties, such as efficiency and asymptotic optimality, thus offering an intuitive interpretation of results (Merlin and Chen 2021).

However, the ordinary least squares method may have some disadvantages in some situations. For example, it may be sensitive to anomalous values, which may significantly influence the estimations of coefficients, rendering the results less reliable. Moreover, when explanatory variables are strongly inter-correlated (multicollinearity problem), the method may become unstable and produce biased or inefficient estimations (Kim et al. 2022a, 2022b).

OLS is used for linear data and homoscedastic residuals, which means that the variance of residuals is constant. It is broadly used in many domains such as economy, social sciences, natural sciences and health sciences. Despite its advantages, it is essential to admit the limits of the ordinary least squares method and to apply it correctly depending on the context and characteristics of available data (Becker and Clancy 2020).

6.3.2. *Ridge regression*

Ridge regression, also known as L2 regularization, is a technique used for the estimation of regression coefficients aiming to reduce overfitting by adding a penalty on the magnitude of coefficients. This method is particularly useful when data are affected by multicollinearity, which means that explanatory variables are strongly inter-correlated. It is used to find the optimal values of regression coefficients that minimize a cost function, including both the sum of squared residuals and the penalty term associated with L2 regularization (El Hami et al. 2015a, 2015b; Zemzami et al. 2020).

Ridge regression modifies the problem of optimization of ordinary least squares by adding a penalty to the sum of squares of coefficients:

$$\min_{\beta} \sum_{i=1}^{n} \left(y_i - \beta_0 - \beta_1 x_{i1} - \dots - \beta_p x_{ip} \right)^2 + \lambda \sum_{j=1}^{p} \beta_j^2 \qquad [6.6]$$

where λ is the regularization parameter, which controls the extent of the penalty. The higher the value of λ, the stronger the penalty, which leads to smaller estimate coefficients.

The estimations of Ridge regression coefficients can be obtained by solving the modified normal equations:

$$\hat{\beta} = \left(X^T X + \lambda \mathrm{I}\right)^{-1} X^T y \qquad [6.7]$$

This formulation adds a small quantity λ to the diagonal of $X^T X$, which stabilizes the inversion of the matrix and reduces the sensitivity to problems of multicollinearity, where I is the pxp identity matrix (Mermi et al. 2024).

The Ridge regression method is widely used when data present multicollinearity, and there is a risk for the standard linear regression model to produce biased or unstable estimations. By adding a penalty to the magnitude of coefficients, Ridge regression is used to regularize the model and enhance its capacity of generalization to test data. It is commonly applied in various domains such as machine learning, finance and computational biology, among others (Zhang et al. 2023).

However, Ridge regression also has limitations. Indeed, the selection of the regularization parameter λ may be difficult, and cross-validation is often required to find the optimal value. Moreover, the penalty on the coefficients may lead to biased estimations if the value of λ is poorly selected. Finally, Ridge regression cannot be used to automatically select the most important variables, which may limit its interpretability in some cases (Rezaei et al. 2023).

6.3.3. *LASSO regression*

LASSO (Least Absolute Shrinkage and Selection Operator) regression is a linear regression technique that adds a regularization penalty to the ordinary least squares (OLS) method to reduce the number of coefficients of the model (El Hami et al. 2015a). This means that, in addition to yielding accurate predictions, LASSO can also automatically select the most important variables for the prediction (Chen et al. 2023).

The objective of the LASSO regression is to minimize the following cost function, known as loss function of the LASSO regression:

$$LASSO(\beta) = \frac{1}{2n}(\|y - X\beta\|_2^2 + \lambda \sum_{j=1}^{p} |\beta_j| \qquad [6.8]$$

where y is the vector of the observed values of the dependent variable, X is the design matrix (each line representing an observation and each column an explanatory variable), β is the vector of coefficients to be estimated, n is the number of observations in the dataset, p is the number of predictors and λ is the regularization parameter that controls the strength of LASSO penalty.

The first part of the cost function ($\|y - X\beta\|_2^2$) is the sum of squared residuals, similar to the ordinary least squares (OLS) method. The second part ($\lambda\sum_{j=1}^{p}|\beta_j|$) is the LASSO penalty, which is proportional to the sum of absolute values of β_j coefficients. The regularization parameter λ controls the magnitude of this penalty, and therefore the parsimony of coefficients.

By adding the LASSO penalty to the cost function, some coefficients β_j can be forced to zero, which is equivalent to excluding the corresponding model variables. This is how LASSO can automatically select the most important variables for prediction, which may be particularly useful when the number of predictors is high compared to the number of observations (Rachmawati et al. 2021).

LASSO has several advantages, particularly its capacity to efficiently manage datasets with a large number of predictors by automatically selecting the most important variables. Moreover, by supporting the parsimony of coefficients, LASSO can improve the model interpretability by identifying the most reliable variables for the prediction (Soh and Zhu 2022).

However, selecting the regularization parameter λ may be difficult, and cross-validation is often required to find the optimal value. Moreover, LASSO tends to arbitrarily select a single predictor from a group of strongly correlated predictors, which may lead to a loss of potentially useful information. Finally, LASSO penalty may lead to biased estimations of coefficients, particularly when the variables are strongly correlated or the number of observations is small (Zhang et al. 2021).

6.3.4. *Elastic Net regression*

Elastic Net regression is a linear regression technique that combines the penalties of Ridge regression and LASSO regression. It is often used when data have problems of multicollinearity and when the selection of variables is expected. This method involves a complex optimization aimed at finding optimal values of regression coefficients, while balancing the reduction of variance and the selection of variables (El Hami et al. 2011). Optimization plays a crucial role in Elastic Net regression by determining the appropriate weights of penalty terms in order to obtain a model that offers both a good capacity of generalization and an enhanced interpretability (Alrajeh et al. 2012).

Elastic Net regression is based on the following cost function, which combines the penalty of both Ridge regression and LASSO regression:

$$ElasticNet(\beta) = \frac{1}{2n}(\|y - X\beta\|_2^2 + \lambda_1 \sum_{j=1}^{p} \beta_j^2 + \lambda_2 \sum_{j=1}^{p} |\beta_j| \quad [6.9]$$

where y is the vector of observed values of the dependent variable, X is the design matrix, β is the vector of coefficients to be estimated, n is the number of observations in the dataset, p is the number of predictors, λ_1 and λ_2 are the regularization parameters that control the extent of the penalty of Ridge regression and of LASSO regression, respectively.

Elastic Net regression offers several significant advantages. First of all, it efficiently manages the problems of multicollinearity, thus reducing the variance of coefficient estimators. This property is inherited from Ridge regression, which confers robustness to Elastic Net when facing datasets that have high correlations between predictors. Furthermore, Elastic Net also integrates the automatic selection of variables, similar to the LASSO regression, which can be used to regularize the coefficients toward zero and select the most reliable variables for prediction. By combining these two aspects, Elastic Net offers a balanced solution, ensuring both the stability of the model and an easier interpretation of results (Han and Dawson 2021).

Despite its many advantages, Elastic Net regression also has some drawbacks. First, the increased complexity of the model can render the process of setting up regularization parameters more difficult compared to more simple methods such as Ridge regression or LASSO regression. Second, even though the automatic selection of variables enhances the model parsimony, it can also render the interpretation more complex, particularly when many variables are involved. Finally, the presence of two regularization parameters to be adjusted may add an additional layer of complexity, requiring an in-depth comprehension of their impact on model performance (Su and Wang 2021).

To summarize, Elastic Net regression is a powerful technique that combines the advantages of Ridge regression and LASSO regression. It can be particularly useful when data have multicollinearity problems and the selection of variables is expected. However, setting up the regularization parameters may be difficult and the interpretation of the model may be more complex in the presence of an automatic selection of variables (Shiomi et al. 2022).

6.3.5. *Weighted least squares*

The weighted least squares method (WLS) is an extension of the ordinary least squares method (OLS), used when data have no constant homoscedasticity, i.e. when model residuals have different variances. Unlike the ordinary least squares method, which assigns the same weight to all observations when estimating the coefficients of the model, the weighted least squares method can be used to assign various weights to each observation depending on the reliability or the importance of these observations (Jang et al. 2023).

Mathematically, the weighted least squares method consists of minimizing the weighted sum of squared residuals, defined as follows:

$$MCP(\beta) = \sum_{i=1}^{n} w_i \left(y_i - X_i \beta \right)^2 \qquad [6.10]$$

where y_i represents the observed value of the dependent variable for the ith observation, X_i is the vector of values of explanatory variables for the ith observation, β is the vector of coefficients to be estimated and w_i is the weight assigned at the ith observation.

There are various ways in which the weights w_i can be defined depending on the study context. For example, they can be inversely proportional to the variance of corresponding residuals, which gives more weight to observations presenting more reliable residuals. Another common approach consists of using weights based on previous knowledge on the reliability of observations, assigning greater weights to observations considered more accurate or more important (Dong et al 2023).

The weighted least squares method has several significant advantages. First, it can be used to take the differences in reliability between observations into account, which may improve the accuracy and robustness of the model. By giving more weight to more reliable observations, the weighted least squares method reduces the influence of less reliable or anomalous observations on the estimations of coefficients. Moreover, this method can also improve the efficiency of model parameter estimation by maximizing the use of available information in the data (Wenxin et al. 2023).

Despite these advantages, the weighted least squares method also has some limitations. First, the appropriate selection of weights may be difficult and often requires an in-depth comprehension of the study context and of the available data. A poor specification of weights may lead to biased or inefficient estimations of the model coefficients. Furthermore, the use of incorrect weights may also affect the predictive performance of the model by introducing a systematic bias in the predictions. Finally, the weighted least squares method may be more complex to implement and interpret compared to the ordinary least squares method, because of the necessity to specify and justify the weights assigned to each observation (Kotha et al. 2023).

6.3.6. *Linear regression by gradient descent*

Linear regression by gradient descent is an iterative optimization approach broadly used for the estimation of coefficients of a regression model. It involves the iterative updating of coefficients to minimize the cost function (Chander and Das 2023). Coefficients are updated using the following updating rule:

$$\beta_j := \beta_j - \alpha \frac{\partial_J(\beta)}{\partial \beta_i} \qquad [6.11]$$

where α is the learning rate (or step), $J(\beta)$ is the cost function (generally the sum of squared errors) and $\frac{\partial_J(\beta)}{\partial \beta_j}$ is the gradient of the cost function with respect to the coefficient β_j.

The cost function J in the linear regression by gradient descent is defined as the sum of squared errors between the observed values y_i and the predicted values $\widehat{y}_i$ for all observations i. Mathematically, it is defined as follows:

$$J(\beta) = \frac{1}{2m} \sum_{i=1}^{m} (y_i - \widehat{y}_i)^2 \qquad [6.12]$$

where m is the number of observations in the dataset and β represents the coefficients of the model.

The basic principle of gradient descent is the iterative calculation of the cost function gradients with respect to each coefficient, followed by the update of coefficients in the direction opposite to gradient to progressively reach the global minimum. This method is particularly efficient for large size datasets or when the cost function is not convex.

Considering the above, the relation becomes:

$$\beta_j := \beta_j - \alpha \frac{1}{m} \sum_{i=1}^{m} (y_i - \hat{y}_i) x_{ij} \qquad [6.13]$$

where x_{ij} is the value of the jth explanatory variable for the ith observation.

Linear regression by gradient descent has several significant advantages. First, it is highly evolutive, which means that it can be applied to large sets of data without requiring an excessive quantity of computational resources. Consequently, it is a method adapted to real problems involving large amounts of data, such as large-scale machine learning or big data analysis. Moreover, this approach is flexible and can be extended to solve nonlinear regression problems, thus offering a multipurpose solution to a wide range of predictive analysis tasks (Katopodis et al. 2022).

However, despite its advantages, linear regression by gradient descent also has some important limitations. First of all, the selection of the learning rate is crucial, and an inappropriate selection may drive an inefficient convergence or even the divergence of the algorithm. Finding a balance between a too low learning rate, which slows down the convergence, and a too high rate, which may cause instability, is a major challenge when applying this method. Moreover, though gradient descent often converges to a local minimum of the cost function, there is always a risk of convergence to a local minimum rather than the expected global minimum, which may compromise the model accuracy. Moreover, linear regression by gradient descent may be sensitive to scale characteristics, which means that data normalization may be needed to guarantee an efficient convergence and accurate results (Paquin et al. 2023).

6.4. Model evaluation

Once the linear regression model is constructed, it is essential to evaluate its performance to ensure its relevance and reliability. This section explores three key aspects of model evaluation: performance evaluation metrics, cross-validation and model diagnosis.

6.4.1. *Performance evaluation metrics*

Performance evaluation metrics are used to quantify the quality of the model and its capacity to explain the variations of the observed data. The most commonly used metrics include those given in the following sections.

6.4.1.1. *Mean square error (MSE)*

The mean square error (MSE) measure is commonly used to evaluate the performance of regression models, including linear regression. It represents the mean of squared gaps between the values predicted by the model and the actual values of the dependent variable in the dataset. MSE is a measure of the mean error of the model, and it is used to compare the accuracy of various models or to evaluate the overall performance of a specific model (Liu et al. 2023a, 2023b). The equation is the following:

$$MSE = \frac{1}{n}\sum_{i=1}^{n}\left(y_i - \hat{y}_i\right)^2 \quad [6.14]$$

where n is the number of observations in the dataset, y_i is the actual value of the dependent variable for observation i and $\hat{y}_i$ is the predicted value of the dependent variable for observation i.

A lower value of MSE indicates a better fitting of the model to data, as this means that model predictions are closer to actual values. In other terms, a low MSE suggests that the model effectively captures the variation of data and succeeds in providing accurate estimations. On the other hand, a high MSE indicates an average fitting of the model to data, which means that model predictions are far from the observed values (Kim et al. 2022a).

MSE is widely used to compare various regression models and select the one providing the best predictive performances. It can also be used to assess the impact of changes brought to model parameters and to data characteristics on its capacity to provide accurate predictions. Finally, MSE is an essential tool for the diagnosis of potential problems of a model, such as underfitting or overfitting, by identifying the situations in which the model does not succeed to correctly capture the data structure. To summarize, the mean square error plays a crucial role in the evaluation and improvement of regression models, providing a quantitative measure of their predictive performance (Karunasingha 2022).

6.4.1.2. *Coefficient of determination (R^2)*

The coefficient of determination, often denoted by R^2, is a statistical measure commonly used to evaluate the quality of the fitting of a regression model to observed data. It represents the proportion of the variance of the dependent variable that is explained by the regression model (Barrett 1974). In other terms, it indicates the extent to which the values predicted by the model correspond to the actual values of the dependent variable:

$$R^2 = 1 - \frac{SSE}{SST} \quad [6.15]$$

where SSE is the sum of squared errors, i.e. the sum of squared gaps between the observed values and the values predicted by the model, and SST is the total sum of squares, i.e. the sum of squared gaps between the observed values and the mean of the observed values. Therefore, the formula becomes:

$$R^2 = 1 - \frac{\sum_{i=1}^{n}(y_i - \hat{y}_i)^2}{\sum_{i=1}^{n}(y_i - \bar{y})^2} \quad [6.16]$$

The coefficient of determination ranges between 0 and 1, where a value of 1 indicates a perfect fitting of the model to data, i.e. the model explains all data variation. By contrast, a value of 0 indicates that the model explains no variation of data and its predictions are as good as those made by simply using the mean of the observed values. Therefore, an R^2 close to 1 indicates that the model effectively captures the variation of data and is able to provide accurate predictions. On the other hand, an R^2 close to 0 suggests that the model does not succeed to explain the variance of data and may require improvement or a review of its specifications (Ozer 1985).

Despite its advantages, the coefficient of determination has some limits that should be taken into account during its use. First of all, the R^2 is sensitive to explanatory variables included in the model. Useless or poorly selected variables can distort the interpretation of the coefficient and lead to erroneous conclusions on the fitting of the model. Moreover, the R^2 is less adapted to evaluate the fitting of nonlinear models, as it is designed to measure the performance of linear models. Finally, it is essential to recognize that the coefficient cannot establish a causal relation between variables. A high R^2 does not necessarily ensure a causal relation between the variables included in the model. Consequently, even though R^2 might be an accurate measure for evaluating the fitting of regression models, it is important to carefully interpret it and add other measures when appropriate (Zhang 2017).

6.4.1.3. *Mean absolute error (MAE)*

The mean absolute error (MAE) is a measure of the accuracy of a prediction model. It is used to evaluate the difference between the values predicted by the model and the actual values observed in a set of data.

The MAE measures the mean of absolute values of gaps between the predictions of the model and the actual observations (Tang et al. 2024).

The MAE is calculated by taking the mean of the absolute values of residuals for each observation i in the set of data, where n is the total number of observations:

$$MAE = \frac{1}{n}\sum_{i=1}^{n}\left|y_i - \widehat{y}_i\right| \qquad [6.17]$$

In this formula, y_i represents the real observed value, $\widehat{y}_i$ represents the value predicted by the model for the observation i and $|\cdot|$ indicates the absolute value of the difference between the real value and the predicted value.

MAE is often used in the context of regression, where it is used for evaluating the accuracy of regression models in terms of prediction of continuous values. For example, in the field of prediction of real estate prices, MAE can be used to measure the accuracy of a regression model that predicts house prices depending on certain characteristics such as size and location (Frías-Paredes et al. 2018).

The main advantage of MAE is the simplicity of its interpretation. Given that it measures the mean of absolute gaps between the predicted values and the real values, it provides a direct indication of the mean accuracy of the model. A lower MAE indicates a better accuracy of the model, while a higher MAE indicates a poorer accuracy (Langdon et al. 2016).

However, MAE does not take the direction of the gaps between the predicted and actual values into account. Consequently, it can be sensitive to anomalous values in the data. Moreover, as it uses the absolute value of gaps, it can be less sensitive to large gaps between predictions and observations, compared to other measures of the error such as the mean square error (MSE). This means that the MAE may underestimate the severity of errors when great errors occur. The MAE is a simple and easy to interpret measure of the accuracy of a regression model, but it may not always be the most appropriate measure depending on the context and on the specific characteristics of data (Coyle and Lin 1988).

6.4.2. *Cross-validation*

Cross-validation is a fundamental technique used in machine learning and statistics to evaluate the performances of predictive models and estimate their capacity of generalization to new data. Its main objective is to reliably estimate the performance of a model on unseen data by effectively using the available set of data. There are several types of cross-validation, each being adapted to specific contexts and needs. The two most commonly used cross-validation techniques are the K-fold cross-validation and the leave-one-out cross-validation (Deng 2023).

6.4.2.1. *K-fold cross-validation*

K-fold cross-validation is a technique for the evaluation of performances of predictive models widely used in machine learning and statistics. It aims to reliably estimate a model capacity of generalization to new data by effectively using the available set of data (Anandan and Manikandan 2023).

The operating principle of k-fold cross validation is the methodical division of the set of data into k subsets, also known as "folds". This division is conducted in such a manner that each fold is successively used as a validation set, while the remaining k-1 folds are combined to form the training set. Hence, each observation of the set of data is used both for training and for the evaluation of the model (Vu et al. 2022).

Once the division has been completed, the model is trained on k-1 folds of the training set. It is then evaluated on the remaining fold, used as a set of validation. This process is repeated k times, each time using a different fold as validation set. Therefore, each observation is used exactly once as a validation set and k-1 times as a training set (Li et al. 2023).

To estimate the performances of the model, an appropriate measure of the error, such as the mean square error (MSE) or the mean absolute error (MAE), is calculated for each iteration of the k-fold cross-validation. The performances obtained for each iteration are then combined by taking the average, which gives a global estimation of the model performance on the set of data (Anandan and Manikandan 2023).

The formula of the mean error for the k-fold cross-validation:

$$CV_k = \frac{1}{k}\sum_{i=1}^{k} Error_i \qquad [6.18]$$

This formula can be used to calculate the mean of errors obtained during each iteration of the cross-validation. Hence, it provides a reliable estimation of the mean performance of the model on the dataset, while taking into account the variability of results due to the division into k-folds (Ling et al. 2019).

K-fold cross-validation has several significant advantages in model performance evaluation. First, it efficiently uses the set of data by dividing them into k-folds and reusing them for model training and evaluation. Due to this exhaustive use of data, a more reliable estimation of the model performances can be obtained. Furthermore, by averaging the performances over several iterations with various validation sets, the k-fold cross-validation provides a more robust evaluation that is less exposed to random fluctuations. Finally, this method reduces the risk of selection bias of the test set, as each sample is used both as part of the validation set and as part of the training set at various moments (Saud et al. 2020).

Though k-fold cross-validation has undeniable advantages, it also has some limitations. First, this method may be calculation-intensive, especially for big sets of data or complex models, as it requires several training and evaluation iterations. Furthermore, the estimated performances can vary depending on the distribution of data into various folds, which may introduce some variability in the results. Moreover, the sensitivity to the selection of the number of folds k is another potential limit, as an inappropriate value of k may lead to a biased estimation of the model performances (Kanti et al. 2023).

6.4.2.2. *Leave-one-out cross-validation (LOOCV)*

Leave-one-out cross-validation (LOOCV) is a model performance evaluation technique widely used in machine learning and statistics. Unlike the k-fold cross-validation that divides the set of data into k subsets, the LOOCV method uses each observation as a validation set by turn, which results in an exhaustive evaluation of model performances (Liu et al. 2023).

The LOOCV process can be described in several stages. First, each observation of the dataset is selected one by one as a validation set, while all the other observations are used as a training set. Then, the model is trained on the training set and evaluated on the observation left aside. This process is repeated for each observation, so that each observation was used once as a validation set. The LOOCV process can be mathematically formalized as follows. Let n be the total number of observations in the set of data and MSE_i the mean squared error obtained using the observation i as validation set (Durdy et al. 2022).

Then, the mean squared error for the set of data is calculated by taking the average of MSE_i on all observations:

$$MSE_{LOOCV} = \frac{1}{n}\sum_{i=1}^{n} MSE_i \quad [6.19]$$

The mean square error of LOOCV is therefore the average of mean square errors calculated for each observation left aside.

The main advantage of LOOCV cross-validation is its robustness and reliability in the estimation of model performances, as it uses all the available observations for the evaluation. Moreover, it avoids the potential bias introduced by the random selection of the number of folds in the k-fold cross-validation. However, LOOCV cross-validation may be calculation-intensive, especially for large sets of data, as it requires model training and evaluation for each observation. Moreover, as each iteration of LOOCV leads to the construction of a model on a slightly different data subset, the variability of results may be higher than that of k-fold cross-validation (Kelter 2021).

6.4.3. *Model diagnosis*

Linear regression model diagnosis is an essential step in the analysis process, aiming to evaluate the fitting quality of the model and to identify the potential problems or biases. It is a crucial step for ensuring the reliability of results and an appropriate interpretation of the coefficients. It can be used to identify the potential problems or biases in the model and take the corrective measures required to enhance its performance and robustness. The most commonly used diagnoses include residual analysis, multicollinearity detection and analysis of influences.

6.4.3.1. *Residual analysis*

Residual analysis consists of assessing the quality of model fitting by examining the distribution of residuals, which represent the differences between the observed values and the values predicted by the model. Ideally, residuals should be random, and follow a normal distribution around zero. Breaches of this condition may indicate errors in model specification or the presence of structures that were not captured by the model (Pagan and Hall 1983).

Residuals are calculated as the difference between the observed values y_i and the values $\widehat{y}_i$ predicted by the model:

$$e_i = y_i - \widehat{y}_i \qquad [6.20]$$

The normality of residuals can be assessed by statistical tests such as Shapiro–Wilk test or by simple visualization of the histogram of residuals.

6.4.3.2. *Multicolinearity detection*

Multicolinearity detection is important as it takes place when two or more explanatory variables are strongly inter-correlated. This may render the coefficient estimations unstable or unreliable. To detect multicollinearity, variance inflation factors (VIF) can be calculated, which measure the extent of multicollinearity for each explanatory variable. High values of VIF suggest a strong multicollinearity and require particular attention (Sundus et al. 2022).

The variance inflation factor (VIF) for each explanatory variable X_j is calculated as follows:

$$VIF\left(X_j\right) = \frac{1}{1 - R_j^2} \quad [6.21]$$

where R_j^2 is the coefficient of determination resulting from the regression of X_j on all the other explanatory variables.

6.4.3.3. *Analysis of influences*

The analysis of influences aims to identify the observations with a disproportionate impact on the results of the model. Influential observations can be detected by examining measures such as studentized residuals, leverages and Cook's distances. Studentized residuals measure the impact of each observation on model fitting, while the leverages measure the influence of each observation on the estimations of coefficients. Cook's distances can be used to detect the observations that would have a significant impact on the model if they were removed (Kr. Das 2015).

The studentized residuals t_i measure the impact of each observation on the model fitting and are calculated as follows:

$$t_i = \frac{e_i}{\sqrt{\hat{\sigma}^2 \left(1 - h_{ii}\right)}} \quad [6.22]$$

where e_i is the residual for the observation, $\hat{\sigma}^2$ is the error variance estimator and h_{ii} is the leverage for observation i.

Leverages h_{ii} measure the influence of each observation on the estimations of coefficients and are calculated as follows:

$$h_{ii} = X_i \left(X^T X\right)^{-1} X_i^T \quad [6.23]$$

where X_i is the i-th line of the design matrix X.

Cook's distances D_i are used to detect the observations that have a disproportionate impact on the model and are calculated as follows:

$$D_i = \frac{\sum_{j=1}^{n}\left(\widehat{y}_{j(i)} - \widehat{y}_j\right)^2}{p.\widehat{\sigma}^2} \quad [6.24]$$

where $\widehat{y}_{j(i)}$ is the prediction of the model when the observation i is removed from the dataset, $\widehat{y}_j$ is the prediction of the model for observation j, p is the number of parameters of the model and $\widehat{\sigma}^2$ is the error variance estimator.

The evaluation of the linear regression model includes the use of performance evaluation metrics, the cross-validation to estimate the capacity of generalization of the model and the model diagnosis to identify any potential problem. These stages guarantee an appropriate and reliable use of the model in various applications.

6.5. Practical cases

As part of our study on linear regression, we will explore a practical exercise focused on predicting car prices. This approach aims to apply the theoretical concepts of linear regression to a concrete situation, in order to better understand how various variables can influence the market price of a vehicle.

6.5.1. *Data preparation and analysis*

Our dataset consists of information on 205 vehicles, distributed in 26 columns, thus offering a complete range of data for exploration. Given this wealth of data, we can study the relation between car prices and a multitude of characteristics, such as engine specifications, vehicle size and other attributes, both qualitative and quantitative. The variables selected for this study include various attributes, from car dimensions to fuel type, which can be used for an in-depth analysis of the factors influencing car prices. Combining numerical and categorical data, we enrich our analysis and linear regression techniques can be used to model the relation between these variables and the vehicle price. Our analysis started by importing the database into Python and by examining its information to understand its structure and content. Then, we conducted an exhaustive verification of the missing values, making sure the data are prepared by removing zero or missing inputs.

6.5.1.1. *Exploratory data analysis*

Exploratory data analysis is crucial for modeling the linear regression, providing essential information on data distribution, their relations and potential abnormalities.

By covering these aspects, the EDA guides the selection of variables and the model construction, ensuring its relevance and reliability.

By examining the distribution of the most frequently registered car brands, we noted that Toyota represents around 15.6% of all the registrations, followed by Nissan with about 8.8%, and Mazda with about 8.3%. Mitsubishi and Honda are, respectively, in fourth and fifth place, each with a frequency of about 6.3%. These results highlight the distribution of car brands, offering a valuable overview of the composition and variety of registered vehicles.

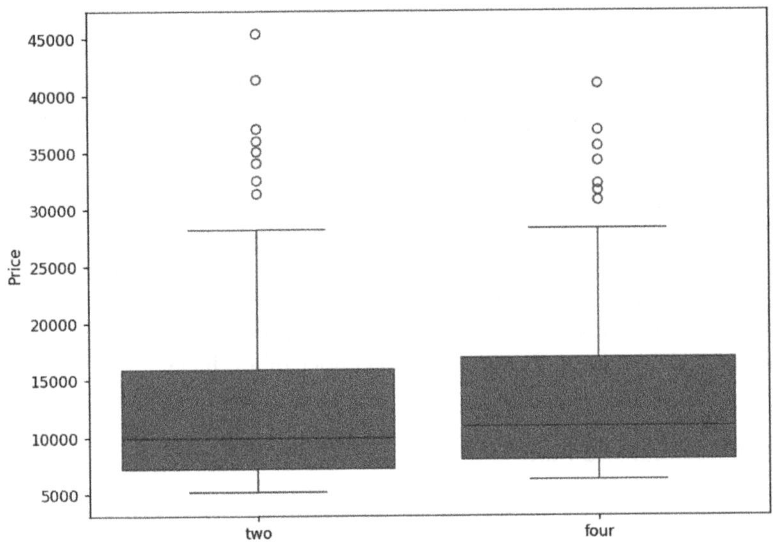

Figure 6.1. *Box diagram for the number of doors*

The statistics extracted from the box diagram (Figure 6.1) for the "doornumber" variable indicate the differences in the distribution of prices as a function of the number of car doors. Four-door cars have a slightly higher median, suggesting higher average prices compared to two-door cars. The anomalous values observed in the two categories indicate the presence of several cases in which prices are significantly higher than the normal for each door configuration. This may reflect factors such as additional functionalities, high-end models or exceptional market conditions for these vehicles.

Globally, these results suggest that the number of doors may play a role in setting car prices, with a tendency toward slightly higher prices for four-door cars.

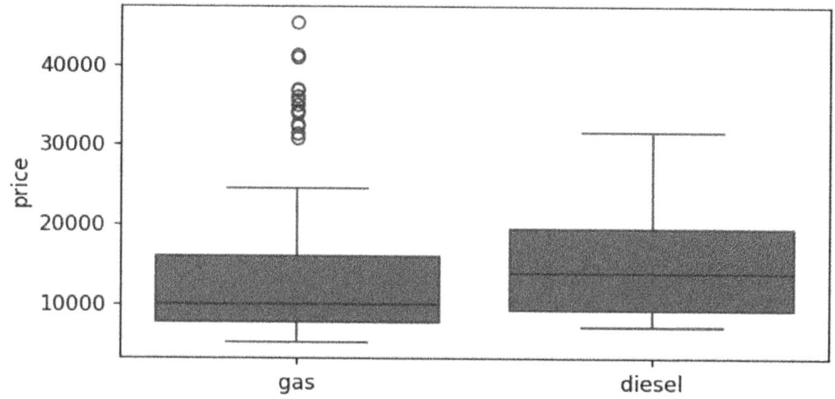

Figure 6.2. *Box diagram for the type of fuel*

Overall, diesel cars appear to be more expensive than gas cars. This observation (Figure 6.2) suggests that the fuel type may significantly influence the car price, diesel cars being generally associated with higher costs. The statistics extracted from the box diagram for the "fueltype" variable confirm this tendency, with a higher median for diesel cars compared to gas cars. Moreover, the absence of anomalous values for diesel cars indicates a more homogeneous distribution of prices, while gas cars have several anomalous values, suggesting a more significant dispersion and very high extreme prices.

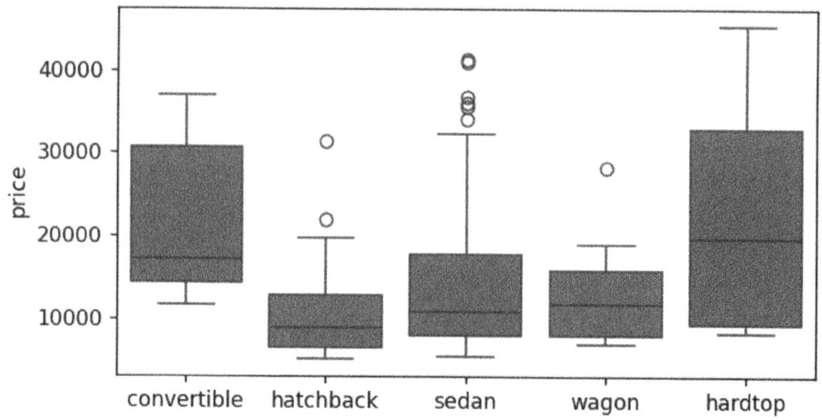

Figure 6.3. *Box diagram for the body type*

The data extracted from the box diagram presented in Figure 6.3 for the "carbody" variable indicate notable disparities in the distribution of prices according to the vehicle body type.

Convertible and hardtop coupe appear to tend toward higher prices, which suggests a market preference for these high-end models. On the other hand, sedan, hatchback and wagon types have relatively lower prices, which may reflect a lower demand and a different perception of the value compared to convertible and coupe models.

These observations highlight the importance of the body type in determining the car prices, with potential implications for pricing and market positioning strategies.

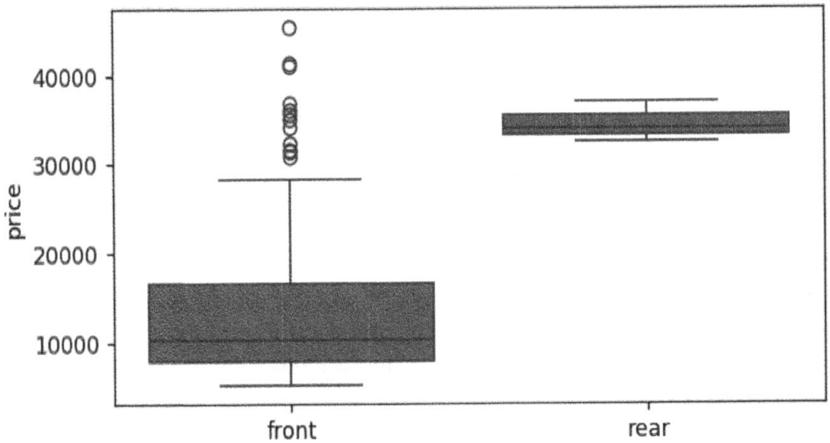

Figure 6.4. *Box diagram for engine location*

According to the statistics extracted from the box diagram in Figure 6.4 for the "enginelocation" variable, there are significant differences in the distribution of prices depending on the car engine location. Vehicles with front engine have a lower median compared to vehicles with rear engine. Lower and higher quartiles for rear engine cars are also higher, suggesting a broader price range.

The absence of anomalous values for rear engine cars indicates consistency in the distribution of prices, while several anomalous values are observed for front engine cars. According to these results, engine location may play a significant role in car pricing, with a tendency to higher prices for rear engine cars.

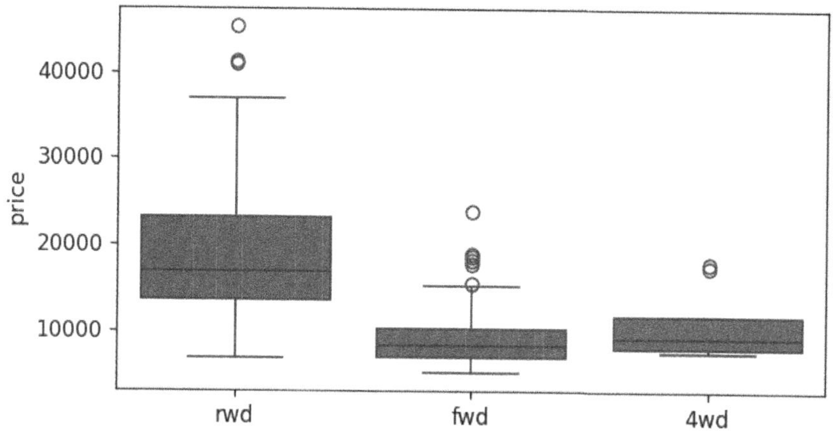

Figure 6.5. *Box diagram for the transmission type*

According to the statistics extracted from the box diagram (Figure 6.5 for the "drivewheel" variable), there are significant variations in the distribution of prices depending on the type of car transmission.

Vehicles with four-wheel drive (4wd) have a lower median compared to vehicles with front wheel drive (fwd) or with rear wheel drive (rwd). Lower and higher quartiles for cars with rear wheel drive are also higher, suggesting a broader price range.

Anomalous values are observed in all transmission categories, indicating the presence of several cases in which prices are significantly above normal for each type of transmission. This may reflect specific characteristics of vehicles, market conditions or other factors influencing the prices.

In summary, these results suggest that the type of transmission may play an important role in car pricing, with various price tendencies depending on the type of transmission.

6.5.1.2. *Data correlation*

The results of correlation (Figure 6.6) between the various characteristics of cars and their price offer precious indications on the relation between these variables. A positive correlation, represented by values close to 1, indicates a tendency in which the value of the characteristic is associated with an increase in the car price.

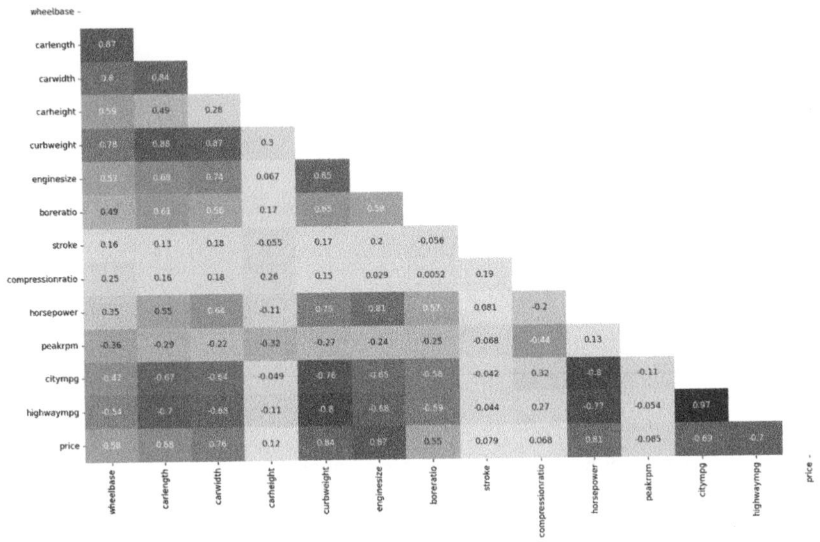

Figure 6.6. *Correlation matrix. For a color version of this figure, see www.iste.co.uk/elhami/applicationsofai.zip*

For example, high positive correlation between "enginesize" and price suggests that cars with larger engines tend to be more expensive. Similarly, characteristics such as "curbweight", "carwidth" and "horsepower" also show significant positive correlations with the price.

On the other hand, negative correlations, indicated by values close to -1, reveal a tendency where an increase in the value of the characteristic is associated with a price reduction. For example, variables such as fuel consumption in the city ("citympg") and on the highway ("highwaympg") have significant negative correlations with the price. This suggests that cars that are more fuel efficient tend to be less expensive, which may be explained by a higher demand for more energy-efficient vehicles.

Finally, weak correlations, represented by values close to zero, indicate a limited linear relation between the variables. For example, the "carheight" shows a relatively weak correlation with the price. This suggests that the car height does not significantly influence its price.

In summary, the analysis of correlations highlights the most important characteristics for predicting car prices. Variables such as the engine size, curb weight and horsepower appear to be significant factors in pricing, while characteristics such as fuel consumption may have a reverse impact on the price. This information is essential for purchasing and pricing decisions in the automotive industry.

6.5.2. *Model training*

6.5.2.1. *Modeling*

For the linear regression modeling aimed at predicting car prices, we deliberately selected a subset of variables featuring a significant correlation with the price, exceeding the threshold of 0.7 in absolute value. The characteristics retained for this analysis are car width, curb weight, engine size, horsepower and fuel consumption on the highway.

These variables have been selected due to their strong correlation with car prices. Indeed, high correlations with the price suggest a stronger linear relation between these characteristics and the price, making them robust potential predictors. Using strongly correlated variables can enhance the accuracy of the linear regression model by grasping more efficiently the car price variations.

By selecting a multiple linear regression, we simultaneously take into account several explanatory variables to predict car prices. In this case, with five explanatory variables, the combined effect of these characteristics on the price can be taken into account in the multiple regression, which can provide a more accurate and complete prediction of the car price than the simple use of a single explanatory variable.

In summary, by selecting variables with high correlations with the price and using a multiple linear regression, we aim to build a robust prediction model to estimate car prices. This approach can be used to fully leverage the relations between the characteristics of cars and their price, by providing valuable information for purchasing and pricing decisions in the automotive industry.

6.5.2.2. *Cross-validation of the model*

In this practical case, we used the k-folds cross validation that, as already seen, is a technique used for evaluating the performances of machine learning models by dividing the dataset into k subsets or folds of equal size. In this case, several specific parameters were used for initialization. First, data were divided into five folds. Each fold is used once as a test set, while the other four folds are combined to form the training set. Several evaluations of the model can be obtained in this way, using various parts of data as test set, which improves the reliability of the evaluation.

Moreover, we indicated that data were mixed before the division into folds. Data mixture is important to avoid any potential bias introduced by the order of data. By mixing data, we make sure that each fold contains a varied representation of classes or characteristics, which yields a more robust and generalizable evaluation.

At each iteration of the loop through the folds of the cross validation, our linear regression model is initialized. The linear regression is used to model the relation between a dependent variable, which in this case is the car price and several independent variables that are the characteristics of the car, such as the width, weight, engine size, horsepower and fuel consumption on the highway. Training aims to fit the model coefficients in order to minimize the error between the predicted values and the actual values of the target, which is the car price in the training set.

6.5.2.3. *Calculation of coefficients and performance evaluation*

Each characteristic (carwidth, curbweight, enginesize, horsepower, highwaympg) is assigned a coefficient to quantify its impact on the car price. At each loop iteration in the linear regression process, coefficients are calculated using the ordinary least squares method (OLS). This involves several steps, particularly the addition of a bias column to the matrix of training characteristics to take into account the ordinate at the origin, the calculation of the transpose of this matrix, its multiplication by itself, the inversion of the resulting matrix and finally the multiplication of this inverse matrix with the product of the transpose of the characteristics matrix and the target vector (car prices). These calculations can be used to obtain the optimal coefficients that minimize the sum of squared residuals, thus providing an accurate estimation of the relations between car characteristics and prices. Table 6.1 presents the results of each fold:

Fold	β_1	β_2	β_3	β_4	β_5	R^2	MAE
1	630.12	1.69	80.21	50.3	-40.43	0.82	2781.27
2	717.49	1.88	86.37	47.1	10.38	0.69	2193.59
3	731.67	2.09	85.55	37.51	5.59	0.77	2237.16
4	334.44	3.5	87.8	55.36	67.15	0.85	2475.26
5	451.51	3.81	67.62	79.04	150.75	0.8	2989.28

Table 6.1. *Simulation results*

– In fold 1, the regression coefficients indicate that the “carwidth” characteristic has the highest coefficient, closely followed by “curbweight”, “enginesize”, “horsepower” and finally “highwaympg”. This suggest that, in this specific fold, the car width is the most influential characteristic for predicting car prices, followed by the curb weight, the engine size, the horsepower and the fuel saving on the highway. The coefficient of determination of 0.82 indicates that the model offers a good explanation of the car price variation in this fold, while the mean absolute error of 2,781 indicates an average accuracy of predictions.

– In fold 2, the regression coefficients show a similar tendency to that of fold 1, with "carwidth" having also the highest coefficient, closely followed by "curbweight", "enginesize", "horsepower" and finally "highwaympg". However, the coefficients are slightly different compared to the previous fold. The coefficient of determination is slightly lower at 0.69, while the mean absolute error slightly decreases to 2193.

– In fold 3, the regression coefficients have similar values to those of previous folds, "carwidth" being still at the top, closely followed by "curbweight", "enginesize", "horsepower" and finally "highwaympg". The coefficient of determination is slightly higher at 0.77, while the mean absolute error remains stable at about 2,237.

– In fold 4, the regression coefficients have a slightly different distribution, with "enginesize", "horsepower" and "highwaympg" having relatively higher coefficients. This suggests that, in the fold, the engine size, the horsepower and the fuel saving on the highway have a stronger influence on car prices. The coefficient of determination has a higher value of 0.85, while the mean absolute error is similar to that of previous folds.

– In fold 5, the regression coefficients feature an increased importance of the characteristic "highwaympg", followed by "horsepower", "curbweight", "enginesize" and finally "carwidth". This suggests that, in this fold, the fuel saving on the highway is the most characteristic for predicting car prices, followed by horsepower, curb weight, engine size and finally the car width. The coefficient of determination is 0.80, while the mean absolute error is slightly higher than in other folds.

The results of simulations across the five folds of cross-validation provide detailed information on the impact of various characteristics on car price prediction. Each fold highlights nuances in the regression coefficients, evidencing the relative importance of characteristics such as car width, curb weight, engine size, horsepower and fuel saving on the highway. Even though in each fold there are variations of the coefficients, fold 4 stands out with a high coefficient of determination of 0.85 and a relatively low mean absolute error of 2475.26, indicating a good capacity of the model to accurately explain the car price variation. However, the final price of the fold will depend on specific objectives of the model and on the preferences in terms of overall performances.

6.5.2.4. *Model diagnosis*

The model diagnosis includes several analyses aimed at evaluating the performance and robustness of the regression model. This section explores three key aspects: analysis of residuals, verification of collinearity and detection of points of influence. First, the analysis of residuals can be used to verify if the differences between the observed and predicted values follow a normal distribution, which would indicate a good fitting of the model to data. Then, the verification of the collinearity aims to identify potential problems of excessive correlation between explanatory variables, which may lead to instability in the estimations of coefficients. Finally, the detection of points of influence consists of identifying the observations with a disproportionate impact on the model, which can be achieved by using tools such as leverages and Cook's distances. A complete overview of the quality and reliability of the regression model can be obtained by combining these analyses.

For fold 1, the residuals, which represent the differences between the observed values and the predicted values, follow a normal distribution, which is a positive sign indicating a good fitting of the model to data. It is however important to note that there is a collinearity between explanatory variables, which may introduce a certain instability in the estimated coefficients. Furthermore, points of influence have been identified, which suggests the presence of atypical observations with a disproportionate impact on the model.

In fold 2, residuals do not follow a normal distribution, which may indicate a less optimal fitting of the model to data. Moreover, the presence of collinearity between the explanatory variables raises concerns related to the interpretation of estimated coefficients. Points of influence are also detected, which suggests that some observations may have a significant impact on model results.

The residuals of fold 3 do not follow a normal distribution either, which raises concerns related to model fitting the data. Similar to the previous folds, the collinearity between explanatory variables is present, which may complicate the interpretation of coefficients. Furthermore, points of influence have been identified, which indicates the presence of atypical observations having a notable impact on the model.

For fold 4, residuals follow a normal distribution, which is a positive aspect indicating a proper fitting of the model. However, the collinearity between explanatory variables persists, which may require an additional investigation to understand the relations between variables. Points of influence are also present, which underlines the importance of verifying atypical observations that might affect model robustness.

The residuals of fold 5 also follow a normal distribution, which is a favorable indication. However, the collinearity between explanatory variables persists, which may challenge the interpretation of coefficients of the model. Moreover, points of influence have been identified, which underlines the need to monitor atypical observations that may have a significant impact on the results of the model.

Overall, the average coefficient of determination for all the folds is 0.785, which suggests that the car price variation is well explained by the model. However, the mean absolute error is 2,535 on average, which indicates an average accuracy of predictions. The presence of collinearity between explanatory variables in all of the folds also suggests that some variables can be redundant or interdependent, which would require a more in-depth analysis. Furthermore, the identification of points of influence highlights the importance of verifying the atypical observations that may affect the reliability of the model.

6.6. Conclusion

Our study confirms the importance and relevance of linear regression in the context of AI and of machine learning prediction. The integration of AI in these modeling processes also offers continuous optimization and improvement opportunities, leading to more accurate predictions and a better comprehension of the studied phenomena.

The analysis of regression coefficients and of model performances through the various folds of cross-validation provided us with detailed information on the impact of variables on car prices prediction. Despite the variations in the coefficients and performances from one fold to another, our model proved its capacity to explain the variation in car prices with an average accuracy, while highlighting the presence of collinearity between explanatory variables and the importance of points of influence.

At the core of our exploration, we have proved how linear regression can be used to predict car prices, by correctly selecting a set of variables strongly correlated with the price. Through multiple linear regressions, we sought to capture the combined effect of these characteristics on the car price, thus offering a more accurate and complete prediction.

This chapter presents an in-depth exploration of the linear regression concept, highlighting its importance and relevance in a context in which nonlinear phenomena are ubiquitous. We examined various coefficient estimation methods, such as the ordinary least squares (OLS) method, Ridge regression, LASSO regression and still others, as well as the analysis of residuals, the detection of multicollinearity and the analysis of influences to ensure model reliability and relevance.

Furthermore, it is essential to admit the limits of the linear regression model, such as its fundamental hypotheses often breached in actual contexts. These limits include nonlinearity, multicollinearity and sensitivity to anomalous values, which may compromise the validity of results and predictions. In some cases, other more sophisticated or better adapted methods may be required for accurately modeling the complex relations between variables.

6.7. References

Abo Al-Kheer, A., Eid, M., Aoues, Y., El Hami, A., Kharmanda, M.G., Mouazen, A.M. (2011). Theoretical analysis of the spatial variability in tillage forces for fatigue analysis of tillage machines. *Journal of Terramechanics*, 48(4), 285–295.

Alrajeh, N., Elmir, B., Bounabat, B., El Hami, N. (2012). Interoperability optimization in healthcare collaboration networks. *Biomediziniche Technik*, 57(5). doi: 10.1515/bmt-2011-0118.

Anandan, B. and Manikandan, M. (2023). Machine learning approach with various regression models for predicting the ultimate tensile strength of the friction stir welded AA 2050-T8 joints by the K-Fold cross-validation method. *Materials Today Communications*, 34, 105286. doi: 10.1016/j.mtcomm.2022.105286.

Barrett, J.W. (1974). The coefficient of determination–some limitations. *The American Statistician*, 28(1), 19–20. doi: 10.1080/00031305.1974.10479056.

Becker, S. and Clancy, R.J. (2020). Robust least squares for quantized data matrices. *Signal Processing*, 176, 107711. doi: 10.1016/j.sigpro.2020.107711.

Ben Abdessalem, A. and El Hami, A. (2015). A probabilistic approach for optimising hydroformed structures using local surrogate models to control failures. *International Journal of Mechanical Sciences*, 96–97, 143–162.

Ben Abdessalem, A., Pagnacco, E., El Hami, A. (2013). Increasing the stability of T-shape tube hydroforming process under stochastic framework. *International Journal of Advanced Manufacturing Technology*, 69(5–8), 1343–1357.

Bouchekourte, M. and El Hami, N. (2022). Optimization of equity allocations of institutional investors: Study of Moroccan case. *International Journal for Simulation and Multidisciplinary Design Optimization*, 13, 12. doi: 10.1051/smdo/2021042.

Bouchekourte, M., Rhouas, S., El Hami, A. (2022). Minimizing the variance of the coverage ratio as an approach to optimize the exchange rate risk of Brent futures contracts. *International Journal for Simulation And Multidisciplinary Design Optimization*, 13, 17. doi: 10.1051/smdo/2022006.

Bouyekhf, R., El Moudni, A., El Hami, A., Zerhouni, N., Ferney, M. (1996). Reduced order modelling of two-time-scale discrete non-linear systems. *Journal of the Franklin Institute*, 333(4), 499–512.

Bouyekhf, R., El Hami, A., El Moudni, A. (2001). Optimal control of a particular class of singularly perturbed nonlinear discrete-time systems. *IEEE Transactions on Automatic Control*, 46(7), 1097–1101.

Chander, G. and Das, S. (2023). Hesitant t-spherical fuzzy linear regression model based decision making approach using gradient descent method. *Engineering Applications of Artificial Intelligence*, 122, 106074. doi: 10.1016/j.engappai.2023.106074.

Chen, S., Li, L., He, L., Xiong, S., Du, N., Chen, H., Hou, L., Zeng, C. (2023). Establishment and validation of a risk assessment model for myopia among Chinese primary school students during the COVID-19 pandemic: A lasso regression approach. *Heliyon*, 9, 10, e20638. doi: 10.1016/j.heliyon.2023.e20638.

Coyle, E. and Lin, J.-H. (1988). Stack filters and the mean absolute error criterion. *IEEE Transactions on Acoustics, Speech, and Signal Processing*, 36(8), 1244–1254. doi: 10.1109/29.1653.

Demirci, B.Ö., Buğdaycı, O., Ertaş, G., Şanlı, D.E.T., Kaya, H., Arıbal, E. (2023). Linear regression modeling based scoring system to reduce benign breast biopsies using multi-parametric US with color Doppler and SWE. *Academic Radiology*, 30, S143–S153. doi: 10.1016/j.acra.2023.01.024.

Deng, A. (2023). Time series cross validation: A theoretical result and finite sample performance. *Economics Letters*, 233, 111369. doi: 10.1016/j.econlet.2023.111369.

Dong, Q., Liu, B., Zhao, H. (2023). Weighted least squares model averaging for accelerated failure time models. *Computational Statistics & Data Analysis (Print)*, 184, 107743. doi: 10.1016/j.csda.2023.107743.

Durdy, S., Gaultois, M.W., Gusev, V.V., Bollegala, D., Rosseinsky, M.J. (2022). Random projections and kernelised leave one cluster out cross validation: Universal baselines and evaluation tools for supervised machine learning of material properties. *Digital Discovery*, 1(6), 763–778. doi: 10.1039/d2dd00039c.

El Attaoui, A., Boukhalfa, A., Rhouas, S., El Hami, N. (2023). Custom application programming interface data extractor applied to the Klarna e-commerce dataset. *Indonesian Journal of Electrical Engineering and Computer Science*, 31(3), 1624. doi: 10.11591/ijeecs.v31.i3.pp1624-1632.

El Hami, A. and Radi, B. (1996). Some decomposition methods in the analysis of repetitive structures. *Computers and Structures*, 58(5), 973–980.

El Hami, A., Lallement, G., Minotti, P., Cogan, S. (1993). Methods that combine finite group theory with component mode synthesis in the analysis of repetitive structures. *Computers and Structures*, 48(6), 975–982.

El Hami, N., Ellaia, R., Itmi, M. (2010). Hybrid evolutionary optimization algorithm MPSO-SA. *International Journal for Simulation and Multidisciplinary Design Optimization*, 4(1), 27–32. doi: 10.1051/ijsmdo/2010004.

El Hami, N., Itmi, M., Ellaia, R. (2011). Reliability-based design and heuristic optimization MPSO-SA of structures. *Advanced Materials Research*, 274, 91–100. doi: 10.4028/www.scientific.net/amr.274.91.

El Hami, N., Itmi, M., El Hami, A. (2015a). Hybrid evolutionary optimization algorithm for structures. *Advanced Materials Research*, 1099, 102109. doi: 10.4028/scientific.net/amr.1099.102.

El Hami, N., Itmi, M., El Hami, A. (2015b). Simulation and optimization of an actuator. *Advanced Materials Research*, 1099, 110–119. doi: 10.4028/scientific.net/amr.1099.110.

Frías-Paredes, L., Mallor, F., Gastón, M., León, T. (2018). Dynamic mean absolute error as new measure for assessing forecasting errors. *Energy Conversion and Management*, 162, 176–188. doi: 10.1016/j.enconman.2018.02.030.

Han, H. and Dawson, K.J. (2021). Applying elastic-net regression to identify the best models predicting changes in civic purpose during the emerging adulthood. *Journal of Adolescence (London, England: Print)*, 93(1), 20–27. doi: 10.1016/j.adolescence.2021.09.011.

Hanley, J.A. (2016). Simple and multiple linear regression: Sample size considerations. *Journal of Clinical Epidemiology*, 79, 112–119. doi: 10.1016/j.jclinepi.2016.05.014.

Huang, C.-H. and Hsieh, S. (2020). Predicting BIM labor cost with random forest and simple linear regression. *Automation in Construction (Print)*, 118, 103280. doi: 10.1016/j.autcon.2020.103280.

Jahandideh, S., Jahandideh, S., Asadabadi, E.B., Askarian, M., Movahedi, M.M., Hosseini, S., Jahandideh, M. (2009). The use of artificial neural networks and multiple linear regression to predict rate of medical waste generation. *Waste Management*, 29(11), 2874–2879. doi: 10.1016/j.wasman.2009.06.027.

Jang, H.J., Shin, S.J., Artemiou, A. (2023). Principal weighted least square support vector machine: An online dimension-reduction tool for binary classification. *Computational Statistics & Data Analysis*, 187, 107818. doi: 10.1016/j.csda.2023.107818.

Kanti, P., Sharma, P., Koneru, B., Banerjee, P., Jayan, K.D. (2023). Thermophysical profile of graphene oxide and MXene hybrid nanofluids for sustainable energy applications: Model prediction with a Bayesian optimized neural network with K-cross fold validation. *FlatChem (Amsterdam)*, 39, 100501. doi: 10.1016/j.flatc.2023.100501.

Karunasingha, D.S.K. (2022). Root mean square error or mean absolute error? Use their ratio as well. *Information Sciences*, 585, 609–629. doi: 10.1016/j.ins.2021.11.036.

Katopodis, R.F., Lima, P.M.V., França, F.M.G. (2022). Functional gradient descent for n-tuple regression. *Neurocomputing (Amsterdam)*, 500, 1016–1028. doi: 10.1016/j.neucom.2022.05.114.

Kelter, R. (2021). Bayesian model selection in theM-open setting – Approximate posterior inference and subsampling for efficient large-scale leave-one-out cross-validation via the difference estimator. *Journal of Mathematical Psychology (Print)*, 100, 102474. doi: 10.1016/j.jmp.2020.102474.

Kharmanda, G., Ibrahim, M.-H., Abo Al-kheer, A., Guerin, F., El Hami, A. (2014). Reliability-based design optimization of shank chisel plough using optimum safety factor strategy. *Computers and Electronics in Agriculture*, 109, 162–171.

Kim, B., Ryu, K.-H., Heo, S. (2022a). Mean squared error criterion for model-based design of experiments with subset selection. *Computers & Chemical Engineering*, 159, 107667. doi: 10.1016/j.compchemeng.2022.107667.

Kim, J., Choi, N., Heo, S. (2022b). An iterative constrained least squares method for continuous piecewise linear approximation. *Computers & Chemical Engineering*, 168, 108056. doi: 10.1016/j.compchemeng.2022.108056.

Kotha, S.K., Rajpathak, B., Mounica, M., Bhuvanagiri, R. (2023). Wide area measurement systems based Power System State Estimation using a Robust Linear-Weighted Least Square method. *Energy Reports*, 9, 23–32. doi: 10.1016/j.egyr.2023.05.046.

Koulou, A., Zemzami, M., El Hami, N., Elmir, A., Hmina, N. (2020). Optimization in collaborative information systems for an enhanced interoperability network. *International Journal for Simulation and Multidisciplinary Design Optimization*, 11, 2. doi: 10.1051/smdo/2019021.

Kr. Das, E.Al.M. (2015). Observations and cutoffs of different influence measures in multiple linear regression. *Journal of Computational & Theoretical Statistics*, 2(2), 79–85. doi: 10.12785/ijcts/020202.

Langdon, W.B., Dolado, J.J., Sarro, F., Harman, M. (2016). Exact mean absolute error of baseline predictor, MARP0. *Information & Software Technology*, 73, 16–18. doi: 10.1016/j.infsof.2016.01.003.

Lee, K., Im, S., Lee, B. (2023). Prediction of renewable energy hosting capacity using multiple linear regression in KEPCO system. *Energy Reports*, 9, 343–347. doi: 10.1016/j.egyr.2023.09.121.

Li, J., Gao, F., Lin, S., Guo, M., Li, Y., Liu, H., Qin, S., Wen, Q.Y. (2023). Quantum k-fold cross-validation for nearest neighbor classification algorithm. *Physica A*, 611, 128435. doi: 10.1016/j.physa.2022.128435.

Ling, H., Qian, C., Kang, W., Liang, C., Chen, H. (2019). Combination of support vector machine and K-fold cross validation to predict compressive strength of concrete in marine environment. *Construction & Building Materials*, 206, 355–363. doi: 10.1016/j.conbuildmat.2019.02.071.

Liu, Y., Zhang, S., Ma, S., Zhang, Q. (2020). Tests for regression coefficients in high dimensional partially linear models. *Statistics & Probability Letters*, 163, 108772. doi: 10.1016/j.spl.2020.108772.

Liu, Y.-X., Heuvelink, G.B.M., Bai, Z., He, P., Jiang, R., Huang, S.-H., Xu, X.-P. (2022). Statistical analysis of nitrogen use efficiency in Northeast China using multiple linear regression and Random Forest. *Journal of Integrative Agriculture*, 21(12), 3637–3657. doi: 10.1016/j.jia.2022.08.054.

Liu, Q., Wang, Z., He, X., Dong, H., Jiang, C. (2023a). An approximate minimum mean-square error estimator for linear discrete time-varying systems: Handling Try-Once-Discard protocol. *Automatica (Oxford)*, 147, 110656. doi: 10.1016/j.automatica.2022.110656.

Liu, X., Yu, M., Jia, Q., Yan, F., Zhou, Y., Wang, Q. (2023b). Leave-one-ion-out cross-validation for assisting in developing robust QSPR models of ionic liquids. *Journal of Molecular Liquids*, 388, 122711. doi: 10.1016/j.molliq.2023.122711.

Mabrouk, I.B. and El Hami, A. (2019). Effect of number of blades on the dynamic behavior of a Darrieus turbine geared transmission system. *Mechanical Systems and Signal Processing*, 121, 562–578.

Makhloufi, A., Aoues, Y., El Hami, A. (2016). Reliability based design optimization of wire bonding in power microelectronic devices. *Microsystem Technologies*, 22(12), 2737–2748.

Merlin, M.L. and Chen, Y. (2021). Analysis of the factors affecting electricity consumption in DR Congo using fully modified ordinary least square (FMOLS), dynamic ordinary least square (DOLS) and canonical cointegrating regression (CCR) estimation approach. *Energy*, 232, 121025. doi: 10.1016/j.energy.2021. 121025.

Mermi, S., Akkuş, Ö., Göktaş, A., Gündüz, N. (2024). A new robust ridge parameter estimator having no outlier and ensuring normality for linear regression model. *Journal of Radiation Research and Applied Sciences*, 17, 1. 100788. doi: 10.1016/j.jrras.2023.100788.

Moro, T., El Hami, A., El Moudni, A. (2002). Reliability analysis of a mechanical contact between deformable solids. *Probabilistic Engineering Mechanics*, 17(3), 227–232,

Olsen, A.A., McLaughlin, J.E., Harpe, S.E. (2020). Using multiple linear regression in pharmacy education scholarship. *Currents in Pharmacy Teaching and Learning*, 12(10), 1258–1268. doi: 10.1016/j.cptl.2020.05.017.

Ozer, D.J. (1985). Correlation and the coefficient of determination. *Psychological Bulletin*, 97(2), 307–315. doi: 10.1037/0033-2909.97.2.307.

Pagan, A. and Hall, A.D. (1983). Diagnostic tests as residual analysis. *Econometric Reviews*, 2(2), 159–218. doi: 10.1080/07311768308800039.

Paquin, A.L., Chaib-Draa, B., Giguère, P. (2023). Stability analysis of stochastic gradient descent for homogeneous neural networks and linear classifiers. *Neural Networks (Print)*, 164, 382–394. doi: 10.1016/j.neunet.2023.04.028.

Pavelescu, F.M. (2015). Considerations on the correlation between student test statistics in cases of simple and multiple linear regressions. *Procedia. Economics and Finance*, 32, 68–76. doi: 10.1016/s2212-5671(15)01366-0.

Poix, J. and Solev, V.N. (2001). Estimation linéaire sans biais pour des modèles de régression avec bruit stationnaire. *Comptes rendus de l'académie des sciences. Série 1, mathématique*, 333(1), 55–60. doi: 10.1016/s0764-4442(01)01949-8.

Prion, S. and Adamson, K.A. (2020). Making sense of methods and measurements: Simple linear regression. *Clinical Simulation in Nursing*, 48, 94–95. doi: 10.1016/j.ecns.2020.07.004.

Rachmawati, R.N., Sari, A.C., Yohanes, Y. (2021). Lasso regression for daily rainfall modeling at Citeko Station, Bogor, Indonesia. *Procedia Computer Science*, 179, 383–390. doi: 10.1016/j.procs.2021.01.020.

Ren, Q. (2022). Seismic acoustic full waveform inversion based on the steepest descent method and simple linear regression analysis. *Journal of Applied Geophysics*, 203, 104686. doi: 10.1016/j.jappgeo.2022.104686.

Rezaei, I., Amirshahi, S.H., Mahbadi, A.A. (2023). Utilizing support vector and kernel ridge regression methods in spectral reconstruction. *Results in Optics*, 11, 100405. doi: 10.1016/j.rio.2023.100405.

Rhouas, S. and El Hami, N. (2022). A packaging industry optimization based on three metaheuristics methods. In *2022 8th International Conference on Optimization and Applications (ICOA)*, Genoa, 1–5. doi: 10.1109/ICOA55659.2022.9934635.

Rhouas, S., Bouchekourte, M., El Hami, A. (2022). Optimization of the impact measurement of market structure on liquidity and volatility. *International Journal for Simulation And Multidisciplinary Design Optimization*, 13, 9. doi: 10.1051/smdo/2021040.

Rhouas, S., Attaoui, A.E., El Hami, N. (2023). Optimization of the prediction performance in the future exchange rate. In *2023 9th International Conference on Optimization and Applications (ICOA),* Abu Dhabi, 1–6. doi: 10.1109/ICOA58279.2023.10308858.

Sangeetha, J. and Alfia, K.J. (2023). Financial stock market forecast using evaluated linear regression-based machine learning technique. *Measurement Sensors*, 100950. doi: 10.1016/j.measen.2023.100950.

Saud, S., Jamil, B., Upadhyay, Y., Irshad, K. (2020). Performance improvement of empirical models for estimation of global solar radiation in India: A k-fold cross-validation approach. *Sustainable Energy Technologies and Assessments (Print)*, 40, 100768. doi: 10.1016/j.seta.2020.100768.

Shiomi, Y., Toriumi, A., Nakamura, H. (2022). International analysis on social and personal determinants of traffic violations and accidents employing logistic regression with elastic net regularization. *IATSS Research*, 46(1), 36–45. doi: 10.1016/j.iatssr.2021.12.004.

Soh, C.G. and Zhu, Y. (2022). A sparse fused group lasso regression model for fourier-transform infrared spectroscopic data with application to purity prediction in olive oil blends. *Chemometrics and Intelligent Laboratory Systems*, 224, 104530. doi: 10.1016/j.chemolab.2022.104530.

Su, M. and Wang, W. (2021). Elastic net penalized quantile regression model. *Journal of Computational and Applied Mathematics*, 392, 113462. doi: 10.1016/j.cam.2021.113462.

Sundus, K., Hammo, B., Al-Zoubi, M.B., Al-Omari, A. (2022). Solving the multicollinearity problem to improve the stability of machine learning algorithms applied to a fully annotated breast cancer dataset. *Informatics in Medicine Unlocked*, 33, 101088. doi: 10.1016/j.imu.2022.101088.

Tang, Y., Shang, L., Zhang, R., Li, J., Fu, H. (2024). Hybrid divergence based on mean absolute scaled error for incipient fault detection. *Engineering Applications of Artificial Intelligence*, 129, 107662. doi: 10.1016/j.engappai.2023.107662.

Vanhoucke, M. and Colin, J. (2016). On the use of multivariate regression methods for longest path calculations from earned value management observations. *Omega*, 61, 127–140. doi: 10.1016/j.omega.2015.07.013.

Vu, H.L., Ng, K.T.W., Richter, A., An, C. (2022). Analysis of input set characteristics and variances on k-fold cross validation for a Recurrent Neural Network model on waste disposal rate estimation. *Journal of Environmental Management*, 311, 114869. doi: 10.1016/j.jenvman.2022.114869.

Wang, Y., Huang, A.Q., Yao, Z., Zhang, Y. (2024). On a class of linear regression methods. *Journal of Complexity (Print)*, 101826. doi: 10.1016/j.jco.2024.101826.

Wenxin, F., Gu, X., Sun, J., Xu, Z. (2023). Learning-based local weighted least squares for algebraic multigrid method. *Journal of Computational Physics (Print)*, 493, 112437. doi: 10.1016/j.jcp.2023.112437.

Zeine, A.T., El Hami, N., El Hami, A., Ouhimmou, S., Ellaia, R. (2017) Multiobjective optimization of trusses using Backtracking Search Algorithm. *Uncertainties and Reliability of Multiphysical Systems*, 17(1). doi: 10.21494/iste.op.2017.0121.

Zemzami, M., El Hami, N., Itmi, M., Hmina, N. (2018). Optimizing interoperability using a new parallel model based on the PSO algorithm. Paper, HAL [Online]. Available at: https://hal.science/hal-02130913/.

Zemzami, M., El Hami, N., Itmi, M., Hmina, N. (2020). A comparative study of three new parallel models based on the PSO algorithm. *International Journal for Simulation and Multidisciplinary Design Optimization*, 11, 5. doi: 10.1051/smdo/2019022.

Zhang, D. (2017). A coefficient of determination for generalized linear models. *The American Statistician*, 71(4), 310–316. doi: 10.1080/00031305.2016.1256839.

Zhang, S., Wu, J., Jia, Y., Wang, Y., Zhang, Y., Duan, Q. (2021). A temporal LASSO regression model for the emergency forecasting of the suspended sediment concentrations in coastal oceans: Accuracy and interpretability. *Engineering Applications of Artificial Intelligence*, 100, 104206. doi: 10.1016/j.engappai.2021.104206.

Zhang, R., Li, Y., Gui, Y. (2023). Prediction of rock blasting induced air overpressure using a self-adaptive weighted kernel ridge regression. *Applied Soft Computing*, 148, 110851. doi: 10.1016/j.asoc.2023.110851.

Zhang, T., Wang, G.A., He, Z., Mukherjee, A. (2024). Service failure monitoring via multivariate multiple linear regression profile schemes with dimensionality reduction. *Decision Support Systems*, 178, 114122, doi: 10.1016/j.dss.2023.114122.

7

Machine Learning and Artificial Intelligence with XGBoost Algorithm for Binary Classification

7.1. Introduction

Artificial intelligence (AI) is a computer science discipline aiming to endow machines with intellectual capacities similar to those of human beings (Joshi 2020; Kühl et al. 2020). AI has at its core machine learning – an essential branch enabling computer systems to learn and improve from experience and data (Joshi 2020; Kühl et al. 2020).

Machine learning relies on the fundamental concept according to which computers can be programmed to learn from data and conduct specific tasks without direct human intervention (Helm et al. 2020). Rather than following programmed instructions, machine learning systems are trained to recognize models and make decisions based on former examples (Kibria et al. 2018; Helm et al. 2020).

Machine learning has a wide range of applications that are omnipresent in our daily life (Helm et al. 2020). From product recommendations on electronic commerce platforms to smartphone facial recognition, passing through search engines and machine translation systems, machine learning forms the basis of many modern technologies (Helm et al. 2020).

Machine learning techniques range from classical methods, such as supervised and unsupervised learning algorithms, to more advanced approaches, such as deep learning based on artificial neural networks (Kibria et al. 2018; Helm et al. 2020;

Chapter written by Hakima REDDAD, Maria ZEMZAMI, Norelislam EL HAMI and Nabil HMINA.

Joshi 2020; Kühl et al. 2020). Due to these techniques, AI systems are able to process massive amounts of data (Kibria et al. 2018; Helm et al. 2020; Joshi 2020; Kühl et al. 2020), extract relevant information and improve their performances in time.

Machine learning algorithms sparked a revolution in many sectors by using data to support decision-making and improve operational efficiency (Helm et al. 2020; Lee and Shin 2020). In finance and banking (Rhouas et al. 2023), machine learning algorithms are essential for identifying credit risks (Machado and Karray 2022), fraud detection (Dornadula and Geetha 2019; Priya and Saradha 2021) and algorithmic transactions (Boukhalfa et al. 2020b; Sivakumar and Om Prakash 2022). For example, banks use machine learning to analyze transaction schemes and client behavior in order to identify potential frauds in real time. In the healthcare sector (Alshammari 2020; Alanazi 2022), machine learning has an essential role in medical imaging for disease diagnosis (Ferdous et al. 2020; Kadem et al. 2023), such as tumor detection in radiological examinations (Sharif et al. 2020), and in the discovery of medications by predicting the interactions between medications and their targets in order to accelerate the drug development process (Rozenblum et al. 2020). Furthermore, predictive analysis is a useful tool in healthcare management by optimizing resource allocation and predicting the results for patients (Mizan and Taghipour 2022).

In engineering, machine learning plays a crucial role in predictive maintenance, resource allocation, process optimization and detection of anomalies, among other applications (Huang et al. 2017). Furthermore, metaheuristic algorithms became powerful tools in solving complex optimization problems in the engineering field (Hakima et al. 2022; Reddad et al. 2022). By combining machine learning and metaheuristic techniques, researchers and applied scientists are able to address various optimization challenges in all engineering disciplines, leading to improved efficiency, cost reduction and higher performances in real-world applications.

Electronic commerce (Liu 2022) and distribution sectors make extensive use of machine learning for various applications. Recommendation systems fed by algorithms such as collaborative filtering and matrix factorization improve client experience by proposing suggestions of personalized products based on their purchase history and preferences (Kumar et al. 2022). Demand forecasting models use historical sales data and external factors to predict future demand, thus optimizing stock management and supply chain operations (Liu 2022). In telecommunications, machine learning algorithms are used to predict client subscription cancellation, optimize the network (Boukhalfa et al. 2020a; Mahmoud and Ismail 2020) and improve service quality, to ensure optimal performances and client satisfaction.

In the field of marketing and publicity (Ngai and Wu 2022), machine learning can be used in advanced analyses for client segmentation, and sentiment analysis in social media data ad click prediction to optimize advertising campaigns and improve return on investment (Ngai and Wu 2022). In manufacturing and industrial fields (Samine et al. 2022, 2023), predictive maintenance algorithms help prevent equipment failures by detecting anomalies and predicting the maintenance needs based on sensor data, thus reducing downtime and operational costs (Sittón-Candanedo et al. 2018; Bertolini et al. 2021). Furthermore, natural language processing techniques are applied to linguistic translation, sentiment analysis and chatbot development in various sectors, thus improving communication and customer service (Houssein et al. 2021).

Autonomous vehicles (Bachute and Subhedar 2021) and robotics benefit significantly from machine learning (Soori et al. 2023), by using artificial vision algorithms for object detection and recognition, reinforcement learning for trajectory planning, and sensor fusion techniques for navigation and obstacle avoidance (Bachute and Subhedar 2021; Pandharipande et al. 2023). Finally, in the energy sector and public services, machine learning algorithms are used for energy load prediction, power grid failure detection and optimization of renewable energy resources (Narciso and Martins 2020; Arumugam et al. 2022; Forootan et al. 2022).

Machine learning is at the core of these advances, using optimization techniques to minimize cost functions during model training.

At the same time, optimization plays a crucial role in the efficient management of computer resources in AI (Sun et al. 2020). With increasingly complex models, optimization aims to maximize efficiency by reducing calculation costs associated with model training and deployment (Sun et al. 2020). Moreover, the optimization of hyperparameters (Yang and Shami 2022), such as learning rates or network architectures, is essential to maximize the predictive performance of the models.

From this perspective, the standard optimization techniques used in machine learning (Sun et al. 2020), such as gradient descent, can be advantageously completed by the power of metaheuristic algorithms (Hakima et al. 2022; Reddad et al. 2022) to solve complex problems. Metaheuristic algorithms, such as simulated annealing (El Hami et al. 2010, 2011a, 2011b; Zemzami et al. 2018a), the genetic algorithms or the swarms of particles (Zemzami et al. 2016, 2018a, 2018b, 2020; Zemzami 2019a, 2019b), offer flexible approaches to exploring wide solution spaces and finding optimal or quasi-optimal configurations of AI models.

In the remaining sections of this chapter, the machine learning XGBoost (Extreme Gradient Boosting) algorithm will be applied to a practical case study focusing on machine failure prediction. We will start by describing the dataset used, offering a detailed description of the key features and the target variable related to the machine failure indicators.

We will then implement the XGBoost algorithm for predictive modeling, by refining the hyperparameters to optimize the performances using cross-validation techniques. Model performances will then be evaluated using measures such as accuracy, precision, return and receiver operating characteristic (ROC) curve. Moreover, the results of the model will be interpreted by analyzing the feature importance and providing explanations on the predictions.

The remaining part of this chapter is organized as follows: section 7.2 explores the state of the art of the XGBoost algorithm, section 7.3 presents the methodology used to develop and validate a predictive model based on XGBoost to solve a binary classification problem, section 7.4 presents the results obtained and the subsequent discussions, and finally, section 7.5 presents the conclusions and proposes paths for future works.

7.2. XGBoost algorithm: state of the art

In the field of machine learning, there are a large number of algorithms that address various challenges related to data analysis and prediction. Standard methods such as logistic regression (Sperandei 2014) and decision trees (Song and Lu 2015) provide solid bases for tasks such as classification and regression, offering interpretable models with well-defined decision limits. Advanced ensemble techniques such as random forests (Speiser et al. 2019; Nassih et al. 2021) and the support vector machines (SVMs) (Salcedo-Sanz et al. 2014) further improve the accuracy of predictions by combining several models or by taking advantage of the complex transformations of the core. Neural networks, such as the convolutional neural networks (CNNs) and recurrent neural networks (RNNs), are powerful deep learning models capable of learning complex models from large volumes of data (Schmidhuber 2015). They excel in tasks such as image recognition, natural language processing and sequential data analysis, offering top performances in various domains.

Among these methods, gradient-boosting algorithms (Natekin and Knoll 2013) gained their place due to their outstanding performances and their versatility. At the forefront, there is XGBoost (Chen and Guestrin 2016; Uzir et al. 2016; Guo et al. 2020; Asselman et al. 2023), an enhanced version of the gradient boosting algorithm that excels in processing structured/tabular data. XGBoost relies on decision tree ensemble, each being sequentially trained to correct the errors made by the previous trees. Its main innovations include efficient construction of trees, control mechanisms to avoid excessive fitting and conducting parallel computing, which renders it evolutionary and well adapted to large sets of data (Chen and Guestrin 2016; Uzir et al. 2016; Guo et al. 2020; Asselman et al. 2023).

7.2.1. *Anatomy of XGBoost algorithm*

The anatomy of XGBoost algorithm includes several key stages that collectively contribute to the training of an optimized tree ensemble model for predictive tasks. The algorithm begins by inputting data, generally a set of data including features (input variables) and a target variable (output variable). Data preprocessing is an essential initial stage, which includes data cleaning to process missing or anomalous values, as well as coding category variables, if needed. The set of data is then divided into two sets, one for training and the other one for testing, to facilitate the evaluation of the model (Chen and Guestrin 2016; Uzir et al. 2016; Guo et al. 2020; Asselman et al. 2023).

Once the data are prepared, the next stage consists of initializing the model parameters. This involves the definition of hyperparameters such as the number of trees (n_estimators), the maximal depth of each tree (max_depth), the learning rate (eta) and the regularization parameters (lambda, alpha) in order to control the complexity of the model (Chen and Guestrin 2016; Uzir et al. 2016; Guo et al. 2020; Asselman et al. 2023). The XGBoost algorithm uses an iterative approach for the construction of trees within a boosting framework (Chen and Guestrin 2016; Uzir et al. 2016; Guo et al. 2020; Asselman et al. 2023). It begins by initializing the booster (trees ensemble) with an initial prediction, often the mean target variable (Chen and Guestrin 2016; Uzir et al. 2016; Guo et al. 2020; Asselman et al. 2023).

The iterative process for tree construction then continues so that for each tree of the ensemble (Chen and Guestrin 2016; Uzir et al. 2016; Guo et al. 2020; Asselman et al. 2023):

– The gradient (first derivative of the loss function) and the hessian (second derivative of the loss function) are calculated for each instance of the dataset.

– A decision tree is constructed to adapt to the negative gradient (residuals), the instances being weighted depending on the calculated hessian.

– Regularization techniques such as tree depth control and the minimum weight of children are applied to manage the tree complexity and prevent overlearning.

– The predictions of each tree are added to the ensemble, scaled by a learning rate in order to control the contribution of each tree to the final model.

During the learning process, model performances are evaluated using a specific evaluation measure, such as logarithmic loss for classification tasks. The algorithm uses stop criteria to determine when to stop the construction of trees, which can be based on reaching a predetermined performance threshold or on reaching a maximum number of trees. Once the stop criteria are met, the final XGBoost model is obtained as the sum of all the individual trees in the ensemble (Chen and Guestrin 2016; Uzir et al. 2016; Guo et al. 2020; Asselman et al. 2023).

Briefly, the XGBoost algorithm relies on the iterative reinforcement of trees to iteratively improve predictions by sequentially adding decision trees to the ensemble. Each tree is constructed to process the residual errors of the model, by carefully managing the model complexity using hyperparameters and regularization techniques. The result is a robust and efficient tree ensemble model, capable of making accurate predictions on new data. This structured approach to model training and optimization is at the origin of the potential and popularity of XGBoost in various machine learning applications (Chen and Guestrin 2016; Uzir et al. 2016; Guo et al. 2020; Asselman et al. 2023).

The flow chart in Figure 7.1 shows the key stages of the XGBoost algorithm operation.

7.2.2. *Main features of XGBoost algorithm*

7.2.2.1. *Gradient-boosting framework*

The gradient-boosting framework is a robust ensemble learning technique that combines the predictions of several reliable models, such as shallow decision trees, to create a powerful predictive model (Chen and Guestrin 2016; Uzir et al. 2016; Guo et al. 2020; Asselman et al. 2023). It operates by iteratively minimizing a loss function specified by the gradient descent, by adjusting the parameters of weak learners at each iteration to correct the errors made by the existing ensemble (Chen and Guestrin 2016; Uzir et al. 2016; Guo et al. 2020; Asselman et al. 2023). The ensemble grows sequentially, each new weak learner being trained to complete the gaps of the current ensemble (Chen and Guestrin 2016; Uzir et al. 2016; Guo et al. 2020; Asselman et al. 2023). A shrinkage parameter or a learning rate controls the contribution of each weak learner, thus contributing to the regularization and prevention of overlearning. Regularization techniques such as tree size reduction and subsampling of features further improve model robustness. The training process continues to the extent that a stop criterion is met, which leads to reaching optimal performances without overlearning (Chen and Guestrin 2016; Uzir et al. 2016; Guo et al. 2020; Asselman et al. 2023). Finally, the ensemble of weak learners is used for making predictions on new data, producing accurate and generalized results for the regression and classification tasks (Chen and Guestrin 2016; Uzir et al. 2016;

Guo et al. 2020; Asselman et al. 2023). The gradient-boosting framework is illustrated in Figure 7.3.

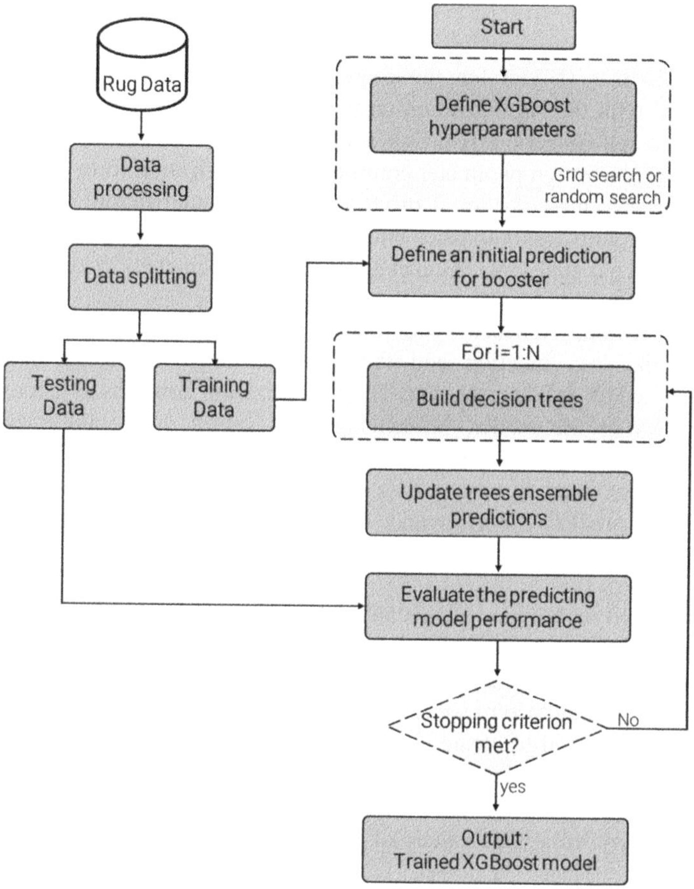

Figure 7.1. *XGBoost algorithm flow chart*

7.2.2.2. XGBoost regularization process

XGBoost involves regularization in its objective function to avoid overlearning and improve the generalization features of the model (Chen and Guestrin 2016; Uzir et al. 2016; Guo et al. 2020; Asselman et al. 2023). The regularization term penalizes models that are too complex by adding a regularization term to the loss function during training (Chen and Guestrin 2016; Uzir et al. 2016; Guo et al. 2020; Asselman et al. 2023).

The global objective function of XGBoost combines the loss function (e.g. the mean square error (MSE) for the regression or binary cross-entropy for the classification) with the regularization term, weighted by a regularization parameter (λ or α) (Chen and Guestrin 2016; Uzir et al. 2016; Guo et al. 2020; Asselman et al. 2023). The regularization parameter controls the regularization force applied during training. Higher values of λ or α drive a stronger regularization, which leads to simpler models with less variance, but with a potentially higher bias (Chen and Guestrin 2016; Uzir et al. 2016; Guo et al. 2020; Asselman et al. 2023).

During learning, XGBoost optimizes the objective function by iteratively minimizing the loss function and the regularization term using optimization techniques based on gradient-descent. The regularization term acts as a constraint on the parameters of the model, influencing their values during optimization. By including regularization in the objective function, XGBoost ensures that the trained models are generalized to unseen data and are less prone to overlearning, which improves their predictive performance in reality (Chen and Guestrin 2016; Uzir et al. 2016; Guo et al. 2020; Asselman et al. 2023).

There are two main types of regularization used in XGBoost, among which L1 regularization (Lasso Regularization) adds a penalty term to the objective function that is proportional to the absolute values of the model parameters. It facilitates dispersion by reducing to zero some coefficients, thus enabling an efficient selection of traits. It can be used to create simpler models and reduce the overlearning risk. The second type of regularization is L2 regularization (Ridge Regularization), which adds a penalty term to the objective function proportional to the square of model parameters. It prevents high values of parameters, which leads to more uniform models whose variance is lower. L2 regularization prevents overlearning by penalizing too complex models (Chen and Guestrin 2016; Uzir et al. 2016; Guo et al. 2020; Asselman et al. 2023).

7.2.2.3. *Parallel processing*

XGBoost is designed to enable efficient parallel processing, in order to be more rapid than the standard gradient-boosting implementations (Chen and Guestrin 2016; Uzir et al. 2016; Guo et al. 2020; Asselman et al. 2023). It is compatible with the parallel processing distributed on a single machine or on a cluster of machines (Chen and Guestrin 2016; Uzir et al. 2016; Guo et al. 2020; Asselman et al. 2023).

XGBoost can be used to leverage parallel processing to simultaneously train decision trees, and thus significantly accelerate the training process (Chen and

Guestrin 2016; Uzir et al. 2016; Guo et al. 2020; Asselman et al. 2023). Each decision tree can be independently constructed, and by distributing each task over several cores or machines, XGBoost can leverage the computing power available in modern material architectures (Chen and Guestrin 2016; Uzir et al. 2016; Guo et al. 2020; Asselman et al. 2023).

XGBoost supports trait parallelism, where various processors or machines work simultaneously on various subsets of traits (Chen and Guestrin 2016; Uzir et al. 2016; Guo et al. 2020; Asselman et al. 2023). With this approach, XGBoost can efficiently process sets of data involving a large number of traits by partitioning the space of traits and by processing each partition independently (Chen and Guestrin 2016; Uzir et al. 2016; Guo et al. 2020; Asselman et al. 2023).

XGBoost is equally in charge with data parallelism, where various processors or machines work on various data subsets (Chen and Guestrin 2016; Uzir et al. 2016; Guo et al. 2020; Asselman et al. 2023). During each learning iteration, XGBoost may distribute the data between several processors or machines, each of them calculating the gradients and updating the parameters independently. These updates are then aggregated to produce the final parameters of the model (Chen and Guestrin 2016; Uzir et al. 2016; Guo et al. 2020; Asselman et al. 2023).

XGBoost parallel processing enables the adaptation to large datasets and to distributed computing environments (Chen and Guestrin 2016; Uzir et al. 2016; Guo et al. 2020; Asselman et al. 2023). By distributing calculations over several cluster nodes, XGBoost is able to process datasets that cannot be memorized by a single machine and to efficiently use the resources in a distributed computing environment (Asselman et al. 2023 Chen and Guestrin 2016; Guo et al. 2020; Uzir et al. 2016).

Parallel processing enhances the efficiency of XGBoost by reducing the overall learning time and resource consumption (Chen and Guestrin 2016; Uzir et al. 2016; Guo et al. 2020; Asselman et al. 2023). Parallel computing helps XGBoost minimize idle times and maximize the use of available computing resources, which accelerates learning and model inference (Chen and Guestrin 2016; Uzir et al. 2016; Guo et al. 2020; Asselman et al. 2023).

7.2.2.4. *Managing missing values*

XGBoost has an embedded mechanism for managing missing values (Figure 7.2) in the datasets (Chen and Guestrin 2016; Uzir et al. 2016; Guo et al. 2020; Asselman et al. 2023). It processes missing values as distinct entities during the learning process, which enables the algorithm to conduct distributions depending on whether the value of a data is missing or not (Chen and Guestrin 2016; Uzir et al. 2016;

Guo et al. 2020; Asselman et al. 2023). Due to this mechanism, XGBoost can make well-founded decisions, even when data are incomplete, which contributes to its robustness and efficiency in real applications (Chen and Guestrin 2016; Uzir et al. 2016; Guo et al. 2020; Asselman et al. 2023).

First, XGBoost processes missing values as a distinct category during the tree construction (Chen and Guestrin 2016; Uzir et al. 2016; Guo et al. 2020; Asselman et al. 2023). It creates distinct decision rules for the instances with missing values for a specific data, making sure that these instances are not excluded from the decision-making process (Chen and Guestrin 2016; Uzir et al. 2016; Guo et al. 2020; Asselman et al. 2023).

Then, during the tree construction process, XGBoost considers both the presence and the absence of missing values as potential points of separation (Chen and Guestrin 2016; Uzir et al. 2016; Guo et al. 2020; Asselman et al. 2023). It assesses the purity gain (e.g. the reduction of impurity or the improvement of the loss function) for each possible distribution, including the distributions based on missing values (Chen and Guestrin 2016; Uzir et al. 2016; Guo et al. 2020; Asselman et al. 2023).

When a sample with a missing value reaches a leaf node of the decision tree, XGBoost assigns it a prediction based on statistics (e.g. the mean or the mode) of samples that reached this node during training (Chen and Guestrin 2016; Uzir et al. 2016; Guo et al. 2020; Asselman et al. 2023). This guarantees that the predictions for the instances with missing values are coherent and significant (Chen and Guestrin 2016; Uzir et al. 2016; Guo et al. 2020; Asselman et al. 2023).

XGBoost optimizes the division decisions to minimize the global loss function, taking into account both the presence and the absence of missing values (Chen and Guestrin 2016; Uzir et al. 2016; Guo et al. 2020; Asselman et al. 2023). It chooses the division that maximizes the gain in purity, taking into account the potential impact of missing values on the division decision (Chen and Guestrin 2016; Uzir et al. 2016; Guo et al. 2020; Asselman et al. 2023).

XGBoost offers a high flexibility in the processing of missing values, enabling users to specify the manner in which missing values should be processed during learning (Chen and Guestrin 2016; Uzir et al. 2016; Guo et al. 2020; Asselman et al. 2023). The options include processing missing values as a separate category, the use of a default direction for the missing values or the specification of customized processing strategies.

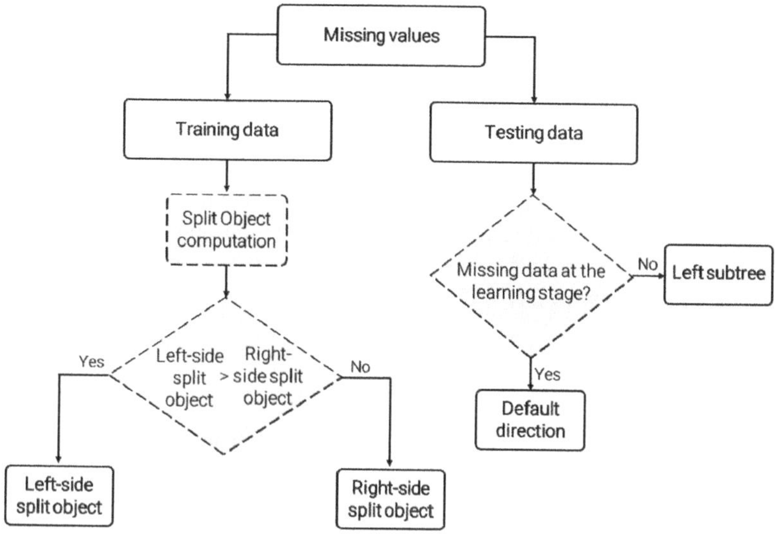

Figure 7.2. *XGBoost missing values processing mechanism*

7.2.2.5. *Tree size reduction*

The algorithm integrates a pruning method that eliminates tree nodes during the training process, which reduces the complexity and prevents overlearning (Chen and Guestrin 2016; Uzir et al. 2016; Guo et al. 2020; Asselman et al. 2023).

Tree size reduction in XGBoost is a technique used to prevent overlearning by eliminating useless nodes or branches in the decision trees during the training process (Chen and Guestrin 2016; Uzir et al. 2016; Guo et al. 2020; Asselman et al. 2023).

7.2.2.6. *Early stopping*

XGBoost implements early stopping as a form of tree pruning (Chen and Guestrin 2016; Uzir et al. 2016; Guo et al. 2020; Asselman et al. 2023). During training, the algorithm permanently monitors the performances on a set of validation data (Chen and Guestrin 2016; Uzir et al. 2016; Guo et al. 2020; Asselman et al. 2023). If the performances on a set of validation data are not improved during a certain number of consecutive iterations (defined by the user by means of a parameter such as "early stopping rounds"), XGBoost stops learning and reduces the trees to optimal iteration, thus preventing all subsequent overlearning (Chen and Guestrin 2016; Uzir et al. 2016; Guo et al. 2020; Asselman et al. 2023).

7.2.2.7. *Maximal depth and minimal weight of children*

XGBoost allows the users to control the maximal depth of trees and the minimal sum of instance (or Hessian) weights required in each leaf node (Chen and Guestrin 2016; Uzir et al. 2016; Guo et al. 2020; Asselman et al. 2023). These parameters act as constraints on the tree structure, preventing trees from becoming too deep or too complex (Chen and Guestrin 2016; Uzir et al. 2016; Guo et al. 2020; Asselman et al. 2023). The trees are reduced if they exceed the maximal depth specified or if the number of instances in a leaf node falls below the minimum threshold of child weight (Chen and Guestrin 2016; Uzir et al. 2016; Guo et al. 2020; Asselman et al. 2023).

7.2.2.8. *Maximum leaf nodes*

Users can also specify the maximum number of leaf nodes authorized in each tree (Chen and Guestrin 2016; Uzir et al. 2016; Guo et al. 2020; Asselman et al. 2023). If a tree exceeds this limit during training, XGBoost reduces the tree by eliminating the leaf nodes or by fusing them with their parent nodes, thus simplifying the tree structure and reducing its complexity (Chen and Guestrin 2016; Uzir et al. 2016; Guo et al. 2020; Asselman et al. 2023).

7.2.2.9. *Gamma (minimal loss reduction)*

XGBoost introduces the “gamma” parameter, also known as minimal loss reduction, which controls the minimal quantity of loss reduction required to conduct an additional distribution on a leaf node (Chen and Guestrin 2016; Uzir et al. 2016; Guo et al. 2020; Asselman et al. 2023).

If the potential gain of the division of a leaf node is below the gamma threshold, XGBoost reduces the node and stops the division, thus avoiding overlearning (Chen and Guestrin 2016; Uzir et al. 2016; Guo et al. 2020; Asselman et al. 2023).

7.2.2.10. *Regularization terms*

The objective function of XGBoost includes regularization terms such as L1 and L2, which penalize too complex models. These regularization terms encourage simpler tree structures by imposing penalties on the high values of parameters, which makes it possible to reduce the tree size during training (Chen and Guestrin 2016; Uzir et al. 2016; Guo et al. 2020; Asselman et al. 2023).

7.2.2.11. *Cross-validation*

XGBoost integrates a cross-validation system to help users find the optimal number of boosting cycles, thus improving model performances.

Cross-validation integrated in XGBoost provides an efficient and automated means to evaluate model performances, select the optimal hyperparameters and estimate the model generalization performances on new data. Consequently, it is possible to construct more robust and reliable models that are less impacted by overlearning and are better generalized on new data (Chen and Guestrin 2016; Uzir et al. 2016; Guo et al. 2020; Asselman et al. 2023).

7.2.3. *XGBoost operation*

The dataset is divided into K layers, generally using a random or sequential division strategy. Each layer contains approximately equal proportions of samples of the set of original data (Chen and Guestrin 2016; Uzir et al. 2016; Guo et al. 2020; Asselman et al. 2023).

XGBoost iterates through each layer, processing a fold as the validation set and the remaining K-1 layers as the learning set (Chen and Guestrin 2016; Uzir et al. 2016; Guo et al. 2020; Asselman et al. 2023). For each layer, XGBoost trains a distinct model using training data and evaluates its performances on validation data (Chen and Guestrin 2016; Uzir et al. 2016; Guo et al. 2020; Asselman et al. 2023).

After training on the K-1 layers, XGBoost evaluates the model performances on the retained fold (validation set). It calculates a performance measure such as the MSE for the regression operations or the logarithmic loss for the classification operations (Chen and Guestrin 2016; Uzir et al. 2016; Guo et al. 2020; Asselman et al. 2023).

XGBoost repeats this process for each layer, which yields K performance measures. The mean of these measures is calculated to obtain an overall estimation of the model performances. Moreover, the users can specify the hyperparameters to set during cross-validation, such as the learning rate, the maximal tree depth and the regularization parameters (Chen and Guestrin 2016; Uzir et al. 2016; Guo et al. 2020; Asselman et al. 2023).

During cross-validation, XGBoost can estimate the optimal number of boosting tours by monitoring the performance measure on the validation set through iterations. It stops the training when the performance measure begins to drop, indicating an overlearning, or when it reaches the maximum number of boosting tours specified by the user (Chen and Guestrin 2016; Uzir et al. 2016; Guo et al. 2020; Asselman et al. 2023).

After cross-validation, XGBoost trains the final model by using the dataset with the optimal hyperparameters determined during cross-validation. This model is then used to make predictions on unseen data (Chen and Guestrin 2016; Uzir et al. 2016; Guo et al. 2020; Asselman et al. 2023).

7.2.4. Theoretical basis of XGBoost algorithm

The XGBoost algorithm leverages the second order Taylor expansion of the loss function and integrates a regularization term to find the optimal solution, which makes it possible to balance the decrease in the objective function and the complexity of the model to avoid overlearning. The XGBoost model is (Chen and Guestrin 2016; Uzir et al. 2016; Guo et al. 2020; Asselman et al. 2023):

$$\widehat{y}_i = \sum_t^N \varphi_t(x_i) \quad [7.1]$$

where N is the number of decision trees, $\varphi_t(x_i)$ is the input function in the t-th decision tree, and $\widehat{y}_i$ is the predicted value (Guo et al. 2020).

The algorithm XGBoost improves as the learning process unfolds, provided that an objective function is defined. The objective function of XGBoost has two components: the loss function, or learning error, and the penalty function (Guo et al. 2020).

The loss function is a method for the evaluation of algorithm performances in terms of dataset modeling (Guo et al. 2020). On the contrary, the penalty function is a regularization term related to model complexity, which prevents algorithm overlearning and supports its generalization performance (Guo et al. 2020).

The objective function of XGBoost algorithm is written as follows:

$$X_{obj} = \sum_{i=1}^n l(y_i, \widehat{y}_i) + \sum_{t=1}^N \Omega(\varphi_t) \quad [7.2]$$

where the sum $\sum_{i=1}^n l(y_i, \widehat{y}_i)$ can be used to calculate the difference between target values and predicted values (Guo et al. 2020). Also, the sum $\sum_{t=1}^N \Omega(\varphi_t)$ and the regularization term, with:

$$\Omega(\varphi_t) = \gamma.T_t + \frac{1}{2}\lambda\|\omega_t\|^2 \quad [7.3]$$

where T_t is the number of leaf nodes of the tree t, ω_t is the leaf node score, γ is the penalty coefficient of leaves, and λ guarantees that the leaf node score is not too high.

The XGBoost algorithm uses a gradient reinforcement; it adds one new tree at a time instead of obtaining all trees at once, and it continuously readjusts the results of previous tests by adjusting the residuals of the last prediction. This strategy can be described as follows (Guo et al. 2020):

$$\begin{cases} \widehat{y}_i^{(0)} = 0 \\ \widehat{y}_i^{(1)} = \varphi_1(x_i) = \widehat{y}_i^{(0)} + \varphi_1(x_i) \\ \widehat{y}_i^{(N)} = \sum_{t=1}^{N} \varphi_t(x_i) = \widehat{y}_i^{(N-1)} + \varphi_N(x_i) \end{cases} \quad [7.4]$$

with formulas [7.2] and [7.3], for the t-th decision tree, the objective function can be updated as follows (Guo et al. 2020):

$$L^t = \sum_{i=1}^{n} l\left(y_i, \widehat{y}_i^{(N-1)} + \varphi_N(x_i)\right) + \Omega(\varphi_t) \quad [7.5]$$

The objective function to be minimized is determined using the second order Taylor expansion of the loss function. The function φ_N is assumed to be sufficiently small to apply the second-degree Taylor formula. Consequently, the objective function is then approximately the following (Guo et al. 2020):

$$L^t = \sum_{i=1}^{n}\left[l\left(y_i, \widehat{y}_i^{(N-1)}\right) + \varphi_N(x_i) g_i + \frac{h_i}{2}\varphi_N^2(x_i)\right] + \Omega(\varphi_t) \quad [7.6]$$

where $g_i = \dfrac{\partial l\left(y_i, \widehat{y}_i^{(N-1)}\right)}{\partial \widehat{y}_i^{(N-1)}}$ is the first derivative (the gradient).

and $h_i = \dfrac{\partial^2 l\left(y_i, \widehat{y}_i^{(N-1)}\right)}{\partial^2 \widehat{y}_i^{(N-1)}}$ is the second derivative (hessian).

Finally, the optimal score and the objective function are expressed as follows (Guo et al. 2020):

$$\omega_j = -\frac{G_j}{H_j + \lambda} \quad [7.7]$$

$$X_{obj} = -\frac{1}{2}\sum_{j=1}^{n}\frac{G_j}{H_j + \lambda} + \lambda T \quad [7.8]$$

where

$$G_j = \sum\nolimits_{i \in I_i} g_i \text{ and } H_j = \sum\nolimits_{i \in I_i} h_i$$

The model constantly calculates the loss of the node during the learning process in order to select the leaf node with the highest gain loss. XGBoost adds new trees by continuously dividing the characteristics. Each time a tree is added, it is in fact a matter of learning a new function $\varphi_N(x_i)$ to adjust the residual of the last prediction. When N trees are obtained after learning, the characteristics of prediction samples will have a corresponding leaf node in each tree and each leaf node corresponds to a score. Finally, the corresponding scores of each tree are added to obtain the prediction value of sample recognition. This process is illustrated in Figure 7.3.

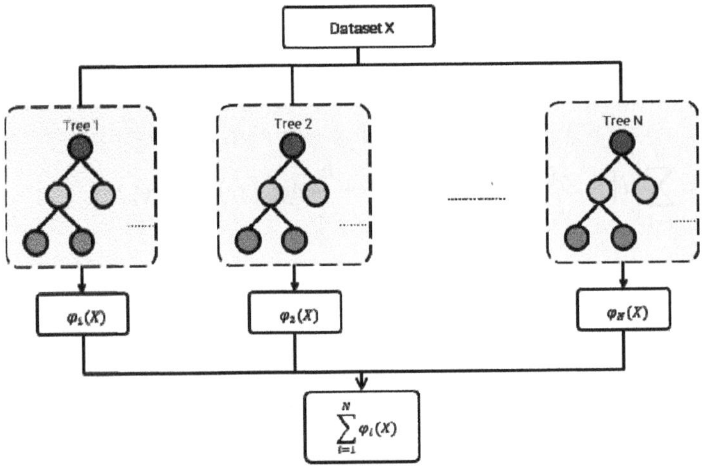

Figure 7.3. *XGBoost operation during iterations. For a color version of this figure, see www.iste.co.uk/elhami/applicationsofai.zip*

7.2.5. Feature importance

The XGBoost algorithm calculates the feature importance scores in order to understand the relative importance of various features in the formulation of predictions (Chen and Guestrin 2016; Uzir et al. 2016; Guo et al. 2020; Asselman et al. 2023). They enable users to identify the features with the most significant impact on model predictions (Chen and Guestrin 2016; Uzir et al. 2016; Guo et al. 2020; Asselman et al. 2023). These scores help users select the features, identify the relevant features and interpret the behavior of the model, which can be used to make better decisions in various machine learning tasks (Chen and Guestrin 2016; Uzir et al. 2016; Guo et al. 2020; Asselman et al. 2023).

XGBoost calculates the feature importance based on the average gain of each feature in all of the decision trees of the ensemble (Chen and Guestrin 2016; Uzir et al. 2016; Guo et al. 2020; Asselman et al. 2023). The gain represents the improvement in the objective function (e.g. the loss reduction) obtained by division on a specific feature. The features with a higher average gain are considered more important as they further contribute to improving the predictive performance of the model (Chen and Guestrin 2016; Uzir et al. 2016; Guo et al. 2020; Asselman et al. 2023).

Another method for calculating the feature importance in XGBoost is based on the number of times that a feature is used to divide data in all the decision trees (Chen and Guestrin 2016; Uzir et al. 2016; Guo et al. 2020; Asselman et al. 2023). The frequently used features for the division are considered as the most important due to their significant impact on the decision-making process of the model (Chen and Guestrin 2016; Uzir et al. 2016; Guo et al. 2020; Asselman et al. 2023).

XGBoost can also calculate the feature importance through the relative coverage of each feature in all decision trees (Chen and Guestrin 2016; Uzir et al. 2016; Guo et al. 2020; Asselman et al. 2023). The coverage measures the proportion of samples passing through the nodes where a specific feature is used for division. Features with higher coverage are considered more important due to their involvement in a larger part of data (Chen and Guestrin 2016; Uzir et al. 2016; Guo et al. 2020; Asselman et al. 2023).

XGBoost normalizes the importance scores of the features to make sure that their sum is equal to 1. This normalization facilitates the comparison of the feature importance between various sets of data and models, providing a consistent measure of the feature importance (Chen and Guestrin 2016; Uzir et al. 2016; Guo et al. 2020; Asselman et al. 2023).

7.3. Case study: application of the XGBoost algorithm – resolution of a binary classification problem

7.3.1. *Problem statement*

The problem of binary classification consists of predicting machine failures based on sensor data and operational conditions, using features such as air temperature, process temperature, rotational speed, torque, tool wear and the type of machine. The dataset includes a binary indicator showing if the machine has failed (1) or not (0) (Narciso and Martins 2020).

The problem is crucial for the implementation of proactive maintenance strategies, which help companies avoid costly failures and optimize operational efficiency. Due to the accurate prediction of the failure probability, companies are able to schedule maintenance interventions in advance, which reduce idle times, minimize repair costs and ensure the continuity of operations.

Machine reliability has a direct impact on production quality, safety and overall performances of the company. By predicting machine failures using advanced analysis and machine learning techniques, companies can make data-based decisions and optimize resource allocation, which improves operational resilience and reduces the risks associated with idle times or with equipment failures.

7.3.2. *Data preparation*

The dataset used for this binary classification problem contains various characteristics providing information on machine operation and state, in order to predict their failures. It mainly consists of the following features:

– ID: a unique identifier assigned to each record, facilitating indexing and referencing.

– Product identifier: it refers to a combination of a machine type variable followed by an identification number, providing an additional context on the monitored specific machine.

– Type: indicates the type of machine for which readings are recorded. The behavior and failure probability of various types of machines may vary according to their operational characteristics.

– Air temperature (Kelvin): ambient temperature around the machine, measured in Kelvin. The variations in ambient temperature may have an impact on machine performances and failure rate.

– Process temperature (Kelvin): the temperature of the process in which the machine is involved, also measured in Kelvin. High process temperatures may stress machine components, which may lead to failures.

– Rotational speed (in revolutions/minute): speed at which the machine operates, measured in revolutions per minute (rev/min). Higher rotational speed may increase the wear of the machine parts, which influences the failure probability.

– Torque (in Nm): measure of the force that rotates the machine, measured in Newton-meters (Nm). High torque levels may indicate significant loads on the machine, which may contribute to potential failures.

– Tool wear (in minutes): this indicates the machine wear level, measured in minutes. A high tool wear indicates that maintenance might be urgently required to avoid potential failures.

– Target variable: machine failure: this refers to the target variable, represented by a binary indicator (0 or 1), indicating whether the machine has failed (1) or not (0). This variable is crucial for the formation of a predictive model that can anticipate machine failures based on recorded characteristics.

7.3.3. Exploratory data analysis

During our exploratory data analysis (EDA), we used various statistical and visual techniques for a better understanding of the dataset.

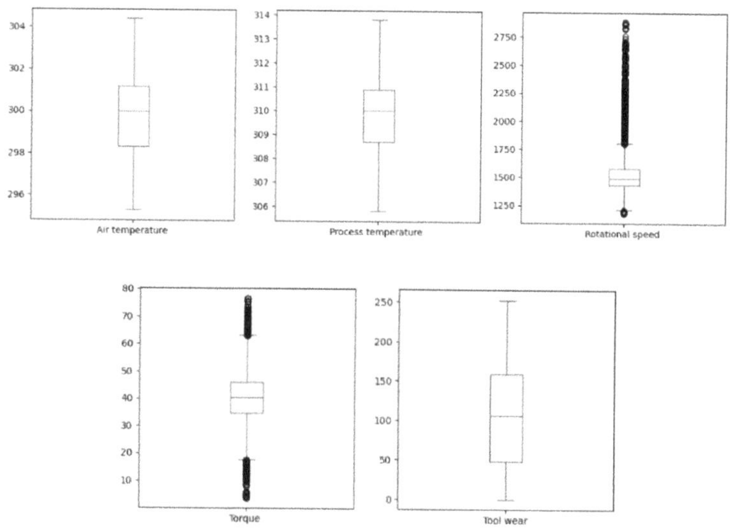

Figure 7.4. *Box diagrams for each feature of the training data*

Box diagrams have been used to visualize the distribution of key characteristics, such as air temperature and rotational speed, highlighting significant variations and potential extreme values susceptible to influencing machine performances. Furthermore, the dispersion diagrams and the histograms have provided an overview on the interactions and distributions of characteristics, offering useful information for subsequent modeling.

7.3.4. *Construction of the predictive model with XGBoost algorithm*

A setting of hyperparameters has been conducted in order to optimize the performance of the XGBoost algorithm for an accurate prediction of machine failures in the binary classification task. By adjusting hyperparameters such as the maximal depth of a tree, the learning rate, the subsampling rate of learning instances, the number of boosting tours (trees), the proportion of subsamples with respect to columns, the minimal sum of instance weight required in a child, the regularization term λ, and γ, the objective being to find an equilibrium between model complexity and generalization. The selected hyperparameters have been finely adjusted to maximize the predictive accuracy while mitigating overlearning. Table 7.1 summarizes the optimal values found and the descriptions of the XGBoost algorithm. The iterative process guarantees that the XGBoost model is optimized for the specific features of the dataset and of the task to be accomplished, which in the end improves its efficiency in the proactive prediction of machine failures.

Hyperparameter	Value	Description
max_depth	13	Controls the maximal depth of each tree in the reinforcement process
Learning_rate	0.017	Determines the size of the step at each iteration by getting closer to optimal weights
colsample_bytree	0.27	Specifies the fraction of features to be randomly sampled for each tree
Subsample	0.92	Defines the fraction of learning data to be used for each boosting tour
n_estimators	530	Is equal to the number of reinforcement cycles (or trees) to be constructed
min_child_weight	6	Specifies the minimal sum of instance weights required in a child node
reg_alpha	0.067	Adds a penalty term to the loss function
reg_lamda	0.39	Uses the L2 norm to penalize significant weights
Gamma	0.035	Is equal to the reduction of minimal loss required to make an additional partition on a leaf node of the tree

Table 7.1. *Optimal values of the hyperparameters of the XGBoost algorithm*

The dataset was divided into training and validation ensembles in a 70–30 proportion, and then the XGBoost model was trained using optimized hyperparameters. This division guarantees that the model learns from a substantial part of data (70%), while saving the remaining part (30%) for the validation in order to evaluate its performances on new data. During the training, the XGBoost algorithm iteratively constructed an ensemble of decision trees, based on specified parameters such as the maximal depth of a tree, the learning rate, the number of boosting tours (trees) and the regularization term λ. The model learned to predict machine failures by minimizing the binary logistics loss function, guided by the objective to maximize the accuracy of the classification. The trained XGBoost model is now ready to be evaluated and deployed to support the decision-making in real time in industrial environments.

7.3.5. *Evaluation of the predictive model*

In this stage, a performance evaluation and an interpretation of the results of the XGBoost model refined for predicting machine failures are conducted. The evaluation measures provide an overview of the predictive accuracy of the model, its capacity to distinguish between positive and negative instances and the importance of features that influence its decisions. By analyzing these measures and by interpreting the model behavior, we aim to obtain information that can be used for proactive maintenance strategies and improve decision-making in industrial environments.

– The accuracy score:

The accuracy score of 0.985 reflects the proportion of correct predictions made by the model on the testing set.

– Confusion matrix:

A confusion matrix is a tool used for the evaluation of performances in machine learning, representing the accuracy of a classification model. It displays a number of true positives, true negatives, false positives and false negatives. It provides a detailed decomposition of the model predictions. In our study, the confusion matrix yielded the following values:

True positives (TPs): 198

False positives (FPs): 138

True negatives (TNs): 40,137

False negatives (FNs): 456.

ROC/AUC score:

The ROC curve and the area under the curve (AUC) score (0.650) evaluate the compromise between the rate of true positives (sensitivity) and the rate of false positives (1 – specificity). An AUC score of 0.650 indicates that the model has higher performances than a random supposition, but it can benefit from an additional optimization, particularly in order to balance sensitivity and specificity.

– Performance analysis:

The XGBoost model reached a high accuracy rate of 98.55%, which indicates a good overall predictive performance. However, a more in-depth analysis using the confusion matrix reveals specific prediction results:

– Accuracy:

$$\frac{TP}{TP+FP}=\frac{198}{198+138}\approx 0.589$$

Accuracy represents the proportion of correctly predicted positive cases (machine failures) among all of the predicted positive instances. In this case, accuracy is moderate, which indicates the possibility to further reduce false positives.

– Return:

$$\frac{TP}{TP+FN}=\frac{198}{198+456}\approx 0.302$$

The return measures the proportion of true positive cases that were correctly predicted by the model. The return rate of about 30.2% suggests that the model may have difficulties in identifying all true positive cases, which indicates potential domains of sensitivity improvement.

– The ROC curve and the AUC score:

The AUC score calculates the area under the ROC curve; it evaluates the compromise between the rate of true positives (sensitivity) and the rate of false positives. A larger AUC indicates a better performance of the model.

The ROC curve with an AUC score of 0.650 illustrates the capacity of the model to distinguish positive instances from negative instances. Even though the AUC score is above 0.5, additional optimizations may be required to enhance the predictive power of the model and the equilibrium between the rates of true and false positives.

7.4. Results and discussions

The refined XGBoost model obtained impressive results in predicting machine failures. The accuracy rate of 98.55% indicates a high level of overall prediction accuracy. Moreover, accuracy and return rates have been calculated at about 58.9% and 30.2%, respectively.

Although the model demonstrates high accuracy, it is possible to improve it by a more efficient identification of true positive cases (machine failures).

On the contrary, the feature importance analysis revealed critical factors influencing machine failures. In particular, rotational speed appeared as the most significant predictive factor, followed by the torque and the air temperature.

These results highlight the importance of monitoring the rotational speed and operational conditions in order to preventively address the maintenance needs.

The setting of hyperparameters posed initial problems due to the complexity of the space of XGBoost parameters. We solved this problem by employing a cross-validation method to identify optimal parameters, by balancing model complexity and performance.

The results of the XGBoost model have significant implications for proactive maintenance strategies in industrial environments.

Based on predictive information on machine failures, companies can implement targeted maintenance interventions, optimize resource allocation and minimize idle times.

Despite the high performance of the XGBoost model, there are limitations, particularly concerning the compromise between accuracy and return. Future research trends include the exploration of ensemble techniques and advanced interpretability methods to improve model interpretability and address the specific challenges of predictive maintenance applications.

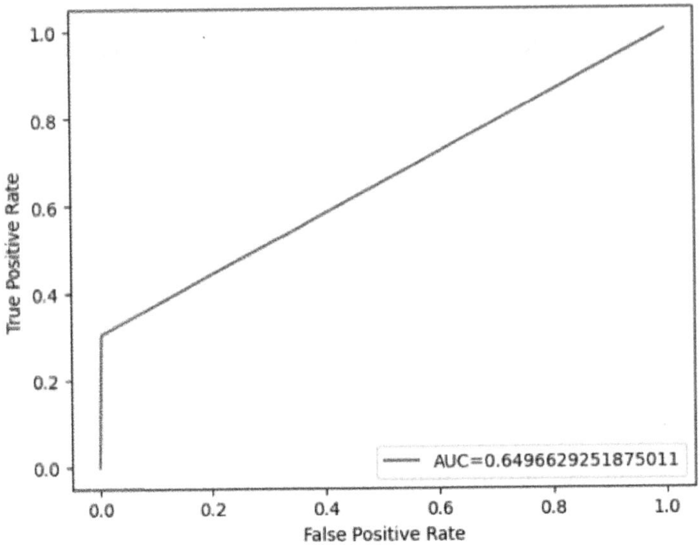

Figure 7.5. *ROC (Receiver Operating Characteristic) curve of XGBoost model*

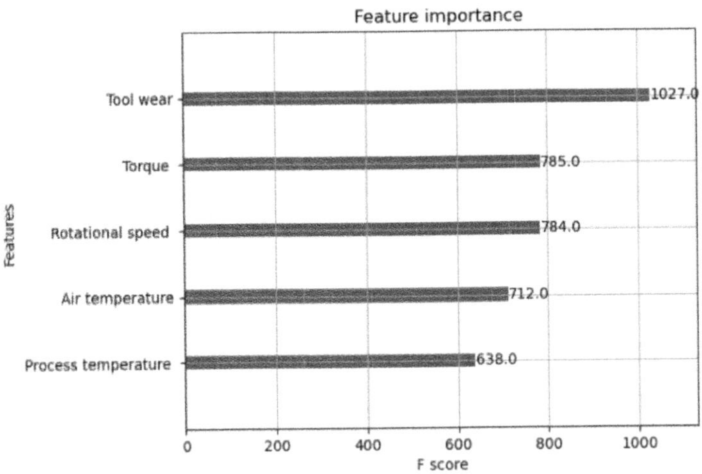

Figure 7.6. *Comparison of feature importance of the binary classification problem*

7.5. Conclusion

In conclusion, this chapter explored the application of machine learning algorithms, with a particular focus on the XGBoost model, for the prediction of machine failures in industrial contexts. As a first step, this chapter presents a state of the art of machine learning algorithms currently used in predictive maintenance, highlighting their theoretical basis and their practical applications.

The case study presented in this chapter demonstrated a systematic approach of predictive modeling, starting with data preprocessing to prepare the dataset for modeling tasks. We applied the XGBoost algorithm, taking advantage of its ensemble of decision trees and regularization techniques to construct a robust predictive model able to identify potential machine failures.

Thanks to model evaluation, we have assessed performances using key measures such as accuracy, precision, return and the ROC curve, which contributed to our better understanding of the predictive capacities of the model. Moreover, we interpreted the results of the model by analyzing the feature importance and by providing explanations for individual predictions, which helped us better understand the underlying factors at the origin of machine failure predictions.

The machine learning domain will continue to evolve in the future, offering new opportunities and new challenges for predictive analysis in various fields. By leveraging the power of advanced algorithms such as XGBoost and integrating them into the domain expertise, organizations can get valuable information and implement proactive strategies to reduce machine failures.

7.6. References

Abo Al-Kheer, A., Eid, M., Aoues, Y., El-Hami, A., Kharmanda, M.G., Mouazen, A.M. (2011). Theoretical analysis of the spatial variability in tillage forces for fatigue analysis of tillage machines. *Journal of Terramechanics*, 48(4), 285–295.

Alanazi, A. (2022). Using machine learning for healthcare challenges and opportunities. *Informatics in Medicine Unlocked*, 30, 2352–9148. doi: 10.1016/j.imu.2022.100924.

Alshammari, F. (2020). Artificial intelligence and machine learning in healthcare applications. *Pharmacophore*, 10(4), 139–149.

Arumugam, K., Swathi, Y., Sanchez, D.T., Mustafa, M., Phoemchalard, C., Phasinam, K., Okoronkwo, E. (2022). Towards applicability of machine learning techniques in agriculture and energy sector. *Materials Today: Proceedings*, 51, 2260–2263. doi: 10.1016/J.MATPR.2021.11.394.

Asselman, A., Khaldi, M., Aammou, S. (2023). Enhancing the prediction of student performance based on the machine learning XGBoost algorithm. *Interactive Learning Environments*, 31(6), 3360–3379. doi: 10.1080/10494820.2021.1928235.

Bachute, M.R. and Subhedar, J.M. (2021). Autonomous driving architectures: Insights of machine learning and deep learning algorithms. *Machine Learning with Applications*, 6, 100164. doi: 10.1016/J.MLWA.2021.100164.

Ben Abdessalem, A. and El Hami, A. (2015). A probabilistic approach for optimising hydroformed structures using local surrogate models to control failures. *International Journal of Mechanical Sciences*, 96–97, 143–162.

Ben Abdessalem, A., Pagnacco, E., El Hami, A. (2013). Increasing the stability of T-shape tube hydroforming process under stochastic framework. *International Journal of Advanced Manufacturing Technology*, 69(5–8), 1343–1357.

Bertolini, M., Mezzogori, D., Neroni, M., Zammori, F. (2021). Machine learning for industrial applications: A comprehensive literature review. *Expert Systems with Applications*, 175, 114820. doi: 10.1016/J.ESWA.2021.114820.

Boukhalfa, A., Abdellaoui, A., Hmina, N., Chaoui, H. (2020a). Network traffic analysis using big data and deep learning techniques. In *6th International Conference on Optimization and Applications, ICOA 2020 – Proceedings*. doi: 10.1109/ICOA49421.2020.9094455.

Boukhalfa, A., Hmina, N., Chaoui, H. (2020b). Parallel processing using big data and machine learning techniques for intrusion detection. *IAES International Journal of Artificial Intelligence (IJ-AI)*, 9(3), 553–560. doi: 10.11591/ijai.v9.i3.pp553-560.

Bouyekhf, R., El Moudni, A., El Hami, A., Zerhouni, N., Ferney, M. (1996). Reduced order modelling of two-time-scale discrete non-linear systems. *Journal of the Franklin Institute*, 333(4), 499–512.

Bouyekhf, R., El Hami, A., El Moudni, A. (2001). Optimal control of a particular class of singularly perturbed nonlinear discrete-time systems. *IEEE Transactions on Automatic Control*, 46(7), 1097–1101.

Chen, T. and Guestrin, C. (2016). XGBoost: A scalable tree boosting system. In *Proceedings of the ACM SIGKDD International Conference on Knowledge Discovery and Data Mining*. ACM, New York, 785–794. doi: 10.1145/2939672.2939785.

Dornadula, V.N, and Geetha, S. (2019). Credit card fraud detection using machine learning algorithms. *Procedia Computer Science*, 165, 631–641. doi: 10.1016/J.PROCS.2020.01.057.

El Hami, A. and Radi, B. (1996). Some decomposition methods in the analysis of repetitive structures. *Computers and Structures*, 58(5), 973–980.

El Hami, A., Lallement, G., Minotti, P., Cogan, S. (1993). Methods that combine finite group theory with component mode synthesis in the analysis of repetitive structures. *Computers and Structures*, 48(6), 975–982.

El Hami, N., Ellaia, R., Itmi, M. (2010). Hybrid evolutionary optimization algorithm MPSO-SA. *International Journal for Simulation and Multidisciplinary Design Optimization*, 4(1), 27–32. doi: 10.1051/IJSMDO/2010004.

El Hami, N., Ellaia, R., Itmi, M. (2011a). A new methodology for the reliability based particle swarm optimization with simulated annealing. *Advanced Materials Research*, 274, 101–111. doi: 10.4028/SCIENTIFIC.NET/AMR.274.101.

El Hami, N., Itmi, M., Ellaia, R. (2011b). Reliability-based design and heuristic optimization MPSO-SA of structures. *Advanced Materials Research*, 274, 91–100. doi: 10.4028/SCIENTIFIC.NET/AMR.274.91.

Ferdous, M., Debnath, J., Chakraborty, N.R. (2020). Machine learning algorithms in healthcare: A literature survey. In *2020 11th International Conference on Computing, Communication and Networking Technologies*. ICCCNT. doi: 10.1109/ICCCNT49239.2020.9225642.

Forootan, M.M., Larki, I., Zahedi, R., Ahmadi, A. (2022). Machine learning and deep learning in energy systems: A review. *Sustainability*, 14(8), 4832. doi: 10.3390/SU14084832.

Guo, R., Zhao, Z., Wang, T., Liu, G., Zhao, J., Gao, D. (2020). Degradation state recognition of piston pump based on ICEEMDAN and XGBoost. *Applied Sciences*, 10(18), 6593. doi: 10.3390/APP10186593.

Hakima, R., Maria, Z., Norelislam, E.H., Nabil, H. (2022). A comparative study of several metaheuristic algorithms for optimization problems. In *8th International Conference on Optimization and Applications, ICOA 2022 – Proceedings*. IEEE.

Helm, J.M., Swiergosz, A.M., Haeberle, H.S., Karnuta, J.M., Schaffer, J.L., Krebs, V.E., Spitzer, A.I., Ramkumar, P.N. (2020). Machine learning and artificial intelligence: Definitions, applications, and future directions. *Current Reviews in Musculoskeletal Medicine*, 13(1), 69–76. doi: 10.1007/S12178-020-09600-8/METRICS.

Houssein, E.H., Mohamed, R.E., Ali, A.A. (2021). Machine learning techniques for biomedical natural language processing: A comprehensive review. *IEEE Access*, 9, 140628–140653. doi: 10.1109/ACCESS.2021.3119621.

Huang, C., El Hami, A., Radi, B. (2017). Metamodel-based inverse method for parameter identification: Elastic–plastic damage model. *Engineering Optimization*, 49(4), 633–653. doi: 10.1080/0305215X.2016.1206537.

Joshi, A.V. (2020). *Machine Learning and Artificial Intelligence*. Springer, Cham. doi: 10.1007/978-3-030-26622-6.

Kadem, M., Garber, L., Abdelkhalek, M., Al-Khazraji, B.K., Keshavarz-Motamed, Z. (2023). Hemodynamic modeling, medical imaging, and machine learning and their applications to cardiovascular interventions. *IEEE Reviews in Biomedical Engineering*, 16, 403–423. doi: 10.1109/RBME.2022.3142058.

Kharmanda, G., Ibrahim, M.-H., Abo Al-kheer, A., Guerin, F., El-Hami, A. (2014). Reliability-based design optimization of shank chisel plough using optimum safety factor strategy. *Computers and Electronics in Agriculture*, 109, 162–171.

Kibria, M.G., Nguyen, K., Villardi, G.P., Zhao, O., Ishizu, K., Kojima, F. (2018). Big data analytics, machine learning, and artificial intelligence in next-generation wireless networks. *IEEE Access*, 6, 32328–38. doi: 10.1109/ACCESS.2018.2837692.

Kühl, N., Goutier, M., Hirt, R., Satzger, G. (2020). Machine learning in artificial intelligence: Towards a common understanding. In *Proceedings of the Annual Hawaii International Conference on System Sciences*, Maui, 5236–5245. doi: 10.24251/hicss.2019.630.

Kumar, B., Roy, S., Sinha, A., Iwendi, C., Strážovská, Ľ. (2022). E-commerce website usability analysis using the association rule mining and machine learning algorithm. *Mathematics*, 11(1), 25. doi: 10.3390/MATH11010025.

Lee, I. and Shin, Y.J. (2020). Machine learning for enterprises: Applications, algorithm selection, and challenges. *Business Horizons*, 63(2), 157–170. doi: 10.1016/J.BUSHOR.2019.10.005.

Liu, L. (2022). E-commerce personalized recommendation based on machine learning technology. *Mobile Information Systems*, 1-2022, 1–11. doi: 10.1155/2022/1761579.

Mabrouk, I.B. and El Hami, A. (2019). Effect of number of blades on the dynamic behavior of a Darrieus turbine geared transmission system. *Mechanical Systems and Signal Processing*, 121, 562–578.

Machado, M.R. and Karray, S. (2022). Assessing credit risk of commercial customers using hybrid machine learning algorithms. *Expert Systems with Applications*, 200, 116889. doi: 10.1016/J.ESWA.2022.116889.

Mahmoud, H.H.H. and Ismail, T. (2020). A review of machine learning use-cases in telecommunication industry in the 5G Era. In *16th International Computer Engineering Conference (ICENCO)*. IEEE.

Makhloufi, A., Aoues, Y., El Hami, A. (2016). Reliability based design optimization of wire bonding in power microelectronic devices. *Microsystem Technologies*, 22(12), 2737–2748.

Mizan, T. and Taghipour, S. (2022). Medical resource allocation planning by integrating machine learning and optimization models. *Artificial Intelligence in Medicine*, 134, 102430. doi: 10.1016/J.ARTMED.2022.102430.

Moro, T., El Hami, A., El Moudni, A. (2002). Reliability analysis of a mechanical contact between deformable solids. *Probabilistic Engineering Mechanics*, 17(3), 227–232.

Narciso, D.A.C. and Martins, F.G. (2020). Application of machine learning tools for energy efficiency in industry: A review. *Energy Reports*, 6, 1181–1199. doi: 10.1016/J.EGYR.2020.04.035.

Nassih, B., Amine, A., Ngadi, M., Azdoud, Y., Naji, D., Hmina, N. (2021). An efficient three-dimensional face recognition system based random forest and geodesic curves. *Computational Geometry*, 97, 101758. doi: 10.1016/J.COMGEO.2021.101758.

Natekin, A. and Knoll, A. (2013). Gradient boosting machines, a tutorial. *Frontiers in Neurorobotics*, 7(DEC), 63623. doi: 10.3389/FNBOT.2013.00021/BIBTEX.

Ngai, E.W.T. and Wu, Y. (2022). Machine learning in marketing: A literature review, conceptual framework, and research agenda. *Journal of Business Research*, 145, 35–48. doi: 10.1016/j.jbusres.2022.02.049.

Pandharipande, A., Cheng, C.H., Dauwels, J., Gurbuz, S.Z., Ibanez-Guzman, J., Li, G., Piazzoni, A., Wang, P., Santra, A. (2023). Sensing and machine learning for automotive perception: A review. *IEEE Sensors Journal*, 23(11), 11097–11115. doi: 10.1109/JSEN.2023.3262134.

Priya, G.J. and Saradha, S. (2021). Fraud detection and prevention using machine learning algorithms: A review. In *Proceedings of the 7th International Conference on Electrical Energy Systems*. IEEE, 564–568. doi: 10.1109/ICEES51510.2021.9383631.

Reade, W. and Chow, A. (2023). Binary classification of machine failures. Competition, Kaggle [Online]. Available at: https://kaggle.Com/Competitions/Playground-Series-S3e17.

Reddad, H., Zemzami, M., El Hami, N., Hmina, N. (2022). Optimisation métaheuristique et en application mécatronique. *Incertitudes et fiabilité des systèmes multiphysiques*, 5(2), 1–13.

Rhouas, S., El Attaoui, A., El Hami, N. (2023). Optimization of the prediction performance in the future exchange rate. In *2023 9th International Conference on Optimization and Applications*. IEEE. doi: 10.1109/ICOA58279.2023.10308858.

Rozenblum, R., Rodriguez-Monguio, R., Volk, L.A., Forsythe, K.J., Myers, S., Mcgurrin, M., Williams, D.H., Bates, D.W., Schiff, G.O., Seoane-Vazquez, E. (2020). Using a machine learning system to identify and prevent medication prescribing errors: A clinical and cost analysis evaluation. *The Joint Commission Journal on Quality and Patient Safety*, 46, 3–10. doi: 10.1016/j.jcjq.2019.09.008.

Salcedo-Sanz, S., Rojo-Álvarez, J.L., Martínez-Ramón, M., Camps-Valls, G. (2014). Support vector machines in engineering: An overview. *WIREs: Data Mining & Knowledge Discovery*, 4(3), 234. doi: 10.1002/widm.1125.

Samine, S., Zemzami, M., Hmina, N., Lagache, M., Belhouideg, S. (2022). Towards the use of artificial intelligence and machine learning in material scientist field. In *8th International Conference on Optimization and Applications*. IEEE. doi: 10.1109/ICOA55659.2022.9934559.

Samine, S., Zemzami, M., Hmina, N., Lagache, M., Belhouideg, S. (2023). Using machine learning and deep learning in additive manufacturing. In *9th International Conference on Optimization and Applications*. IEEE. doi: 10.1109/ICOA58279.2023.10308810.

Schmidhuber, J. (2015). Deep learning in neural networks: An overview. *Neural Networks*, 61, 85–117. doi: 10.1016/J.NEUNET.2014.09.003.

Sharif, M.I., Li, J.P., Naz, J., Rashid, I. (2020). A comprehensive review on multi-organs tumor detection based on machine learning. *Pattern Recognition Letters*, 131, 30–37. doi: 10.1016/J.PATREC.2019.12.006.

Singh, K., Booma, P.M., Eaganathan, U. (2020). E-commerce system for sale prediction using machine learning technique. *Journal of Physics: Conference Series*, 1712(1), 012042. doi: 10.1088/1742-6596/1712/1/012042.

Sittón-Candanedo, I., Nieves, E.H., González, S.R., Martín, M.T.S., Briones, A.G. (2018). Machine learning predictive model for industry 4.0. In *Knowledge Management in Organizations. Communications in Computer and Information Science*, Uden, L., Hadzima, B., Ting, I.H. (eds). Springer, Cham. doi: 10.1007/978-3-319-95204-8_42.

Sivakumar, M. and Om Prakash, P.G. (2022). Machine learning-based algorithmic approach for enhanced anomaly detection in financial transactions. *Lecture Notes on Data Engineering and Communications Technologies*, 93, 779–790. doi: 10.1007/978-981-16-6605-6_59.

Song, Y.Y. and Lu, Y. (2015). Decision tree methods: Applications for classification and prediction. *Shanghai Archives of Psychiatry*, 27(2), 130. doi: 10.11919/J.ISSN.1002-0829.215044.

Soori, M., Arezoo, B., Dastres, R. (2023). Artificial intelligence, machine learning and deep learning in advanced robotics, a review. *Cognitive Robotics*, 3, 54–70. doi: 10.1016/J.COGR.2023.04.001.

Speiser, J.L., Miller, M.E., Tooze, J., Ip, E. (2019). A comparison of random forest variable selection methods for classification prediction modeling. *Expert Systems with Applications*, 134, 93–101. doi: 10.1016/J.ESWA.2019.05.028.

Sperandei, S. (2014). Understanding logistic regression analysis. *Biochemia Medica*, 24(1), 12–18. doi: 10.11613/BM.2014.003.

Sun, S., Cao, Z., Zhu, H., Zhao, J. (2020). A survey of optimization methods from a machine learning perspective. *IEEE Transactions on Cybernetics*, 50(8), 3668–81. doi: 10.1109/TCYB.2019.2950779.

Uzir, N., Raman, S., Banerjee, S., Ramraj S. (2016). Experimenting XGBoost algorithm for prediction and classification of different datasets. *International Journal of Control Theory and Applications*, 9(40), 651–662.

Yang, L. and Shami, A. (2022). On hyperparameter optimization of machine learning algorithms: theory and practice. *Neurocomputing*, 415, 295–316.

Zemzami, M. (2019a). Interoperability optimization using a modified PSO algorithm. *International Journal of Advanced Trends in Computer Science and Engineering*, 8(2), 101–107.

Zemzami, M. (2019b). Variations sur PSO : approches parallèles, jeux de voisinages et applications. Research Thesis, Normandie Université, Ecole nationale des sciences appliquées, Kenitra.

Zemzami, M., El Hami, N., Itmi, M., Hmina, N. (2016). A new parallel approach for the exploitation of the search space based on PSO algorithm. *Colloquium in Information Science and Technology*, 104–110. doi: 10.1109/CIST.2016.7805024.

Zemzami, M., El Hami, N., Itmi, M., Hmina, N. (2018a). An efficient hybrid model based on particle swarm optimization and simulated annealing algorithms for complex optimization problems. *International Journal of Advanced Trends in Computer Science and Engineering*, 8(2), 126–133. doi: 10.30534/ijatcse/2019/05822019.

Zemzami, M., El Hami, N., Itmi, M., Hmina, N. (2018b). Optimizing interoperability using a new parallel model based on the PSO algorithm. *International Journal of Advanced Trends in Computer Science and Engineering*, 8(2), 101–107. doi: 10.30534/ijatcse/2019/01822019.

Zemzami, M., El Hami, N., Itmi, M., Hmina, N. (2020). A comparative study of three new parallel models based on the PSO algorithm. *International Journal for Simulation and Multidisciplinary Design Optimization*, 11, 5. doi: 10.1051/SMDO/2019022.

8

Application of Interoperability to Intelligent System of Systems

Intelligent System of Systems (SoS) draw the attention of the research community as their applications solved many problems of modern society. Recent works identified a large number of challenges related to the various phases of the development cycle of intelligent SoS. One of the most important challenges concerns the evaluation of structural operability. The works in this field are at an early stage of development and lack consistent focus.

The following sections provide a detailed approach to the evaluation of the structural operability of intelligent SoS. This evaluation is conducted through interoperability evaluations taking into account structural dynamics. This aspect is closely related to a specific feature of intelligent SoS, namely the scalability. This is due to their evolution and continuous change.

The remaining part of this chapter is organized as follows:

– Section 1 presents a proposal of definitions, properties and taxonomy of the systems of systems.

– Section 2 presents the concept of interoperability.

– Section 3 explains in detail the metrics of evaluation of interoperability as a basis for the structural analysis of intelligent SoS.

– Section 4 presents theory applications to case studies.

– The final section draws conclusions.

Chapter written by Younes KOULOU, Norelislam EL HAMI, Anas EL ATTAOUI and Sara RHOUAS.

8.1. General context of SoS

8.1.1. *Definitions*

As scientists focused their attention on SoS in recent years, many definitions have been proposed for this concept. Some of the many proposed definitions of SoS are listed as follows:

– "SoS is a set of systems, each being able to operate independently, which interoperate to reach the expected additional abilities" (DoD 2008).

– "SoS are independent systems, operating independently and together to reach higher performance" (Maier 1998).

– "SoS are large-scale integrated systems that are heterogeneous and can operate independently but are in a network for a common goal" (Maier 1998).

– "A SoS consists of multiple, heterogeneous, distributed systems that can operate independently, integrated in multiple level networks that evolve in time" (Lane and Valerdi 2011).

– "SoS is a set of systems able to reach the objective that a single system cannot reach. Each system can operate independently to reach its own objective. SoS has emergence properties" (El Hami et al. 2015a).

– "SoS are special systems, they are composed of systems that can operate independently and have their own advantages and value. Once the elementary system is assembled in a SoS, it maintains its independence and the interactions between systems are frequent" (Xia et al. 2016).

– "SoS are a set of systems working together to reach a more complex target or a higher objective" (Xia et al. 2016).

Although the term SoS was coined quite a while ago, it appears that the concept is still challenging. Tannahill and Jamshidi (2014) convincingly cited, in the statement of Sage and Cuppan (2001), that there was no universally accepted definition of the term "SoS". Moreover, most of the definitions of SoS are not very useful, and some of them are even harmful (Tannahill and Jamshidi 2014).

SoS has a wide range of fields of application. These include software such as the Internet, cloud computing, healthcare and cyber-physical systems, and also hardware-focused fields, such as defense, energy and transportation (DoD 2008).

8.1.2. *SoS taxonomy*

Knowledge transfer to areas involving systems that exclusively have some features of SoS is facilitated by a classification. The proposed classification is based on the management centralization level and on the freedom to operate systems.

According to Madni and Sievers (2014) and to the Systems Engineering Guide for SoS (SEGS) published in 2008, an SoS can be classified based on how it is managed, and also on its ability to adapt to changes.

8.1.2.1. *Directed SoS*

Directed SoS have well-defined objectives and are built to address specific objectives. They may also have assigned responsible, resources, etc. They are managed in a centralized manner during long-term operations to meet the objectives. However, component system (CS) remains autonomous and preserves its capacity to operate independently (e.g. airports).

As illustrated in Figure 8.1, there is a hierarchy of targets among directed SoS. However, the most important objectives are system-strategy objectives that represent global and final objectives of building the strategy system. They are the sought-for solutions of which the entire infrastructure is assembled.

Under the ultimate objectives of SoS, there is a set of objectives that collaborate in order to reach all of the objectives of directed SoS. They form a certain top-down hierarchy of objectives.

The idea behind this hierarchy is to decompose a complex target into several complex objectives. The same process applies to objectives that are also decomposed into several tasks.

The decomposition of targets and objectives into smaller and clearer objectives that are easier to manage and direct will end when we obtain more or less simple and defined tasks that will be assigned to CS.

Consequently, all autonomous and heterogeneous CS will have independent tasks that implicitly contribute to reaching global objectives of SoS.

It is important to note that the notions of "target", "subtarget", "obj" and "task" illustrated in Figures 8.1 and 8.2 aim to underline the hierarchy and demonstrate that there are various objective levels. Therefore, they are used only for explanatory purposes.

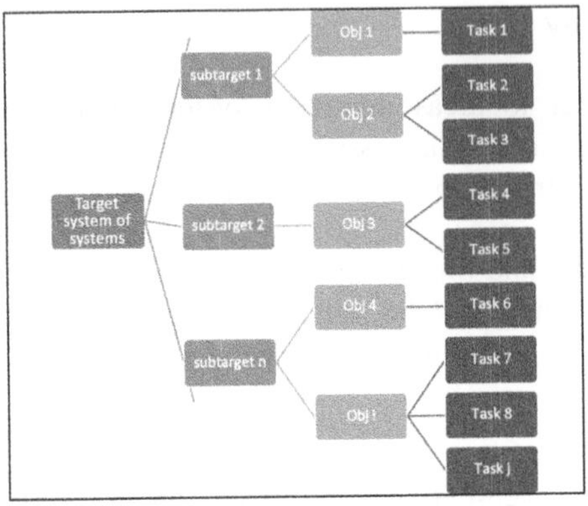

Figure 8.1. *Illustration of directed SoS. For a color version of this figure, see www.iste.co.uk/elhami/applicationsofai.zip*

8.1.2.2. *Open SoS*

In this class of SoS, there can be no central management authority or agreed objective in a centralized manner. However, there are only several objectives on which CS interact more or less voluntarily. They can integrate or leave the management system in a dynamic manner according to the demands of the mission. This class is more exposed to emergence (e.g. national economy).

Unlike directed SoS, open SoS has no central management authority or an objective agreed in a centralized manner. CS participates in a dynamic manner to the performance of SoS in order to reach decentralized objectives.

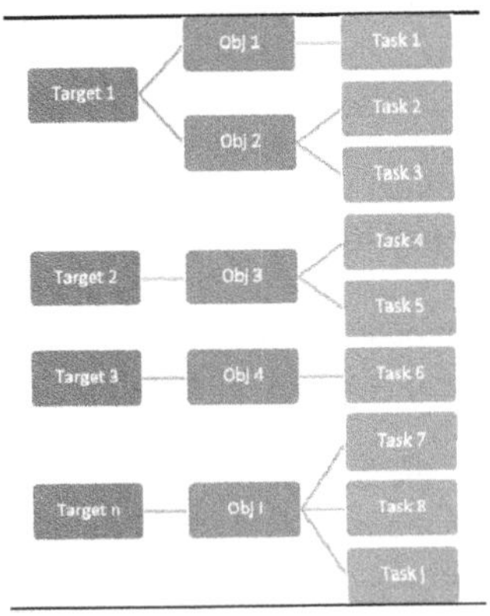

Figure 8.2. *Illustration of open SoS. For a color version of this figure, see www.iste.co.uk/elhami/applicationsofai.zip*

Table 8.1 highlights the main differences between open SoS and directed SoS. It focuses on the differences in management and in the organization of objectives.

Directed SoS	Open SoS
– Centralized management – Well-defined objective(s) – Construction to achieve a specific goal – Possibility of having an appointed manager and resources	– No central management authority – No centrally agreed goal – Voluntary interaction to achieve certain objectives – Dynamic integration and segregation according to mission demands

Table 8.1. *The main differences between open and directed SoS*

It is important to mention that the successful integration of a CS in an SoS, irrespective of its type, requires the definition of the system according to the SoS interoperability rules (Abbott et al. 2006).

8.1.3. *Properties of SoS*

There are qualitative and structural differences between SoS and standard systems, as SoS is not a broader version of the hierarchical structure (Bradshaw et al. 2004). Moreover, they have many properties that differentiate them. These properties are categorized into two classes: systemic properties and functional properties.

8.1.3.1. *Systemic properties of SoS*

At this level, attention is directed towards the CS that forms the SoS and the relations between them. Consequently, the idea is to determine how this structure is constructed. It is important to note that the structure is not stable: it changes and evolves continuously according to the evolution of SoS targets (Bradshaw et al. 2004).

The structure includes the interconnected CS and the resources that support the SoS. The CS may be heterogeneous, autonomous, distributed, interdependent, complex and collaborative.

From a systemic point of view, there is a broad range of properties and features that differentiate SoS from the standard systems.

These properties are detailed as follows.

Autonomy

From a technical point of view, there are two notions involving the concept of autonomy. One of them is direct autonomy, which refers to the independence of a system from any external intervention for proper operation. The other one is self-sufficiency, which refers to the non-dependence of a system on any external intervention to meet its needs (Bradshaw et al. 2003; Boardman and Sauser 2006; Johnson 2014).

Autonomy can be considered as a task for the system itself, as the system can be run independently. However, autonomy involves certain constraints. It should be noted that the constraints should not overwhelm or break its performances (Boardman and Sauser 2006). It is true that a random CS may fail to carry out its task within SoS, but autonomy should not be accepted as a reason. This may be due to a lack of efficiency or even of compatibility.

Boardman and Sauser (2006) state that the notion of autonomy has been neglected in the systemic approach, which explains why some systems are recognized as such, even when they act as elements. In some cases, the difference between an element and a system is difficult to prove, as an element may have relations, operate in a dependent manner and constitute a whole. However, a system is more complex than that, it is a set of elements.

Heterogeneity

Heterogeneity is a very complex issue since SoS should handle the diversity of CS natures in addition to their operation at various time scales. This challenges SoS to act without being affected by the diverging nature and operating schedule of its components.

An SoS must be diversified in terms of resources, functionalities and abilities. Correspondingly, a difference should be made between the SoS focused on demands, based on the defined intent, and the SoS based on the abilities, which have a diversity of functions (Boardman and Sauser 2006).

Moreover, a simple system may have an aperiodic and complex performance, with sometimes infinite trajectory variants that should finally converge to uniform models. This implies that diversity should be handled with caution.

Consequently, the SoS must be heterogeneous in terms of resources and diversity in terms of functionalities and abilities (Boardman and Sauser 2006).

Interdependence

When designing an SoS, it is necessary to create, manage and improve the interdependences and achieve the interoperability of the existing CS.

Interdependences are concerned by the capacity of CS to share and correctly interoperate at the level of information, hardware and sometimes even energy, in order to reach a common objective in a given context following the rules of interaction (DeLaurentis 2005; Billaud et al. 2015). This involves the resolution and the management of CS in an SoS besides their inherent connectivity that does not appear.

Moreover, the systems themselves are responsible for determining their interdependences. This is favorable to the autonomy of self-directedness of systems. However, it is essential to let the full accomplishment of the SoS mission be the guide.

It is important to note that talking about interdependences involves the discussion of interoperability, which represents the capacity of at least two systems to understand one another, share and leverage their functionalities and their information despite their heterogeneity.

Distribution

There is no need to have the CS in the same SoS physically installed in the same geographical location and centralized in terms of management to reach a common objective. Correspondingly, the geographical extent of SoS is wide and vague (Maier 1998). Information and tasks are distributed in the SoS according to certain rules. And this distribution should not be considered as a limitation. Moreover, there is also a management decentralization, which means that the management of the SoS should not rely on a system or a set of systems.

Scalability

There is no fixed structure for SoS. The structure can evolve, extend or even shrink at any time (Bradshaw et al. 2004). This differentiates them from standard systems, as they permanently allow for the integration and segregation of CS.

However, changes should in no way impede reaching the global objective. Correspondingly, the evolutionary model is not exclusive to the structure. In some cases, evolution is related to the environment including the SoS itself.

8.1.3.2. *Functional properties of SoS*

Dealing with an SoS involves facing two realities. The first one is the physical structure representing the CS. The second one is the contribution of this structure, which is the objective of gathering them and making them work together.

Besides the systemic properties, which essentially focus on the structure, there is another important set of properties that characterizes the SoS, called functional properties. At this level, the focus is on the functions of SoS in addition to organization and services.

Besides structure, there are also services that represent defined objectives in terms of collaboration of certain resources in the SoS structure. Services are the object of cooperation and therefore contribute to the progress towards an underlying and global intention. The latter is the object of SoS that represents the sought-for solution of which the entire infrastructure is assembled.

8.2. Concept of intelligent SoS

8.2.1. *Intelligence in SoS*

Intelligence in SoS is a fundamental concept that includes the capacity of SoS to perceive, understand, decide and act in an autonomous and adaptive manner in complex and dynamic environments. This intelligence relies on the efficient collaboration and coordination of various systems that compose the SoS and also on the capacity of the SoS itself to process and use the information provided by its components to make informed decisions (Mahdi 2020).

In an SoS, each individual system can be equipped with its own cognitive and decision-making capacities, but it is the manner in which these capacities are organized and integrated in the SoS that determines its overall intelligence. For example, in a military SoS, the various monitoring, communication and combat systems can operate autonomously to collect information and make local decisions, but it is the capacity of SoS to aggregate and analyze this information to elaborate global strategies that determines its operational efficiency.

Intelligence in SoS also involves the capacity to adapt to changes in the environment and to internal perturbations. SoS must be able to detect changes, react rapidly and consequently modify their behavior to maintain their objectives and their global performance. This often requires the use of machine learning techniques and optimization to improve decision-making and resource management in the SoS.

However, intelligence in the SoS is a multifaceted concept that includes coordination, cooperation, adaptability and the ability to efficiently process information to reach common objectives in complex and dynamic environments. Its efficient development and leverage are essential to maximize the performance and resilience of SoS in a variety of fields of application.

Artificial intelligence (AI) plays a crucial role in the development and operation of SoS. By integrating advanced AI techniques, SoS can improve their ability to collect, analyze and use large amounts of data from multiple sources to make informed decisions and coordinate the actions of various systems that compose them (Mahdi 2020). Several key aspects of how AI is used in SoS are listed as follows.

Machine learning and optimization

SoS can use machine learning algorithms to analyze models and tendencies in the gathered data, and consequently anticipate changes and adapt their strategies.

Autonomous and decision-making systems

SoS that use AI can develop autonomous systems able to make decisions in real time and coordinate their actions without direct human intervention.

Monitoring and detection of anomalies

AI techniques can be used to continuously monitor the performances and health state of the individual systems in an SoS, thus rapidly detecting anomalies or potential failures. Consequently, SoS can take corrective measures before the problems become critical.

Coordination and collaboration

AI can facilitate coordination and collaboration between the various systems in an SoS providing advanced communication and planning mechanisms.

Adaptability and resilience

Using continuous-learning techniques, the SoS can dynamically adapt to changes in the environment and internal perturbations, thus reinforcing their resilience and their capacity to maintain their objectives even under changing or unexpected conditions.

AI offers a significant potential to improve the performances and flexibility of SoS in a variety of fields of application. Its efficient integration in the SoS requires, however, a deep comprehension of their specific needs and constraints, as well as unique challenges related to the management of complex and interconnected systems.

8.2.2. *Architecture of intelligent SoS*

The architecture of intelligent SoS is designed to facilitate a harmonious integration and efficient collaboration between various autonomous CS composing the SoS. This architecture relies on several key principles aiming to ensure the flexibility, scalability and resilience of the system as a whole. We propose an architecture of intelligent SoS illustrated in Figure 8.3, which represents the various layers of architecture of intelligent SoS, from the user interface layer to the interconnection and communication layer, passing through the data collection, analysis and intelligence layers and the control and coordination layers. Each layer is in charge with specific functions that contribute to the overall performance of intelligent SoS.

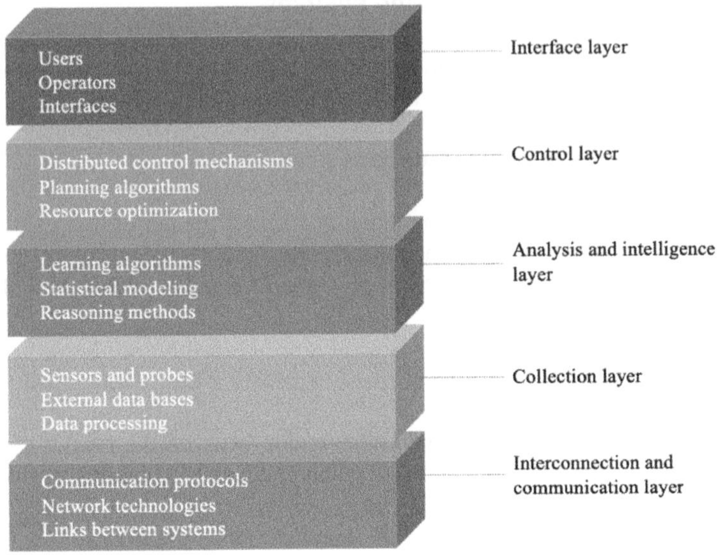

Figure 8.3. *Architecture of intelligent SoS. For a color version of this figure, see www.iste.co.uk/elhami/applicationsofai.zip*

Interconnection and communication layer

This layer is responsible for establishing links and communication between various systems that compose the SoS. It uses standardized-communication protocols and advanced technologies to ensure a reliable and secure connectivity between the systems, leading to a fluid exchange of information and commands.

Data collection layer

This layer is in charge of collecting data from various sources in the SoS, such as sensors, monitoring systems and external databases. It uses advanced-data processing techniques to aggregate, clean and normalize data, thus ensuring their quality and relevance for subsequent decision-making processes.

Analysis and intelligence layer

This layer is in charge with the analysis of collected data to extract significant information and make informed decisions. It uses machine learning algorithms, statistical modeling techniques and advanced reasoning methods to identify models, tendencies and anomalies in the data, thus enabling SoS to proactively respond to changes in the environment and operational needs.

Control and coordination layer

This layer coordinates the actions of various systems in the SoS to reach predefined objectives. It uses distributed-control mechanisms and advanced-planning algorithms to distribute tasks, solve conflicts and optimize the use of available resources, thus ensuring an efficient and harmonious execution of SoS operations.

User interface layer

This layer provides a user-friendly and intuitive interface to human operators who monitor and control SoS activities. It uses advanced-data visualization and human–machine interaction techniques to present relevant information in a clear and comprehensible manner, thus facilitating decision-making and collaboration between humans and automated systems.

These various layers constitute the basic architecture of intelligent SoS, offering an organized and modular structure to design, develop and deploy adaptive and resilient SoS in a variety of fields of application.

8.3. Interoperability

Generally speaking, interoperability is designed as the ability of two systems to understand one another and use their functionalities. The word "interoperate" implies that a system performs an operation for another one. From the perspective of information technology, two heterogeneous computer systems have the ability to jointly operate and share their resources (Chen et al. 2008).

According to the literature, there are various definitions of interoperability that were formulated by various researchers in various sectors and domains. We cite the first two definitions that clarify the concept of interoperability and were formulated in the '90s. According to the first definition, it is "the ability of two or more systems or components to exchange information and to use the exchanged information". The second definition is "the ability – promoted but not guaranteed – is reached thanks to the compliance with a given set of standards, which enable heterogeneous equipment, generally built by different suppliers, to operate together in a network environment" (IEEE 1990).

In the context of our works, we retain the definition specifically adapted to SoS that defines interoperability as:

> An ability of natively independent systems to interact in order to establish harmonious and complete collective behaviors without deep modifications to their structure or to their individual behavior (Pingaud 2009).

8.3.1. *Classification of interoperability*

Interoperability is a quality that can be perceived from many perspectives. This depends on the levels of implementation: the approaches to implementation or the barriers limiting interoperability. There is also the scope of application, the field of concern or the transactional character of the cooperation. Hence, an illustrative classification of interoperability along four axes is proposed, as illustrated in Figure 8.4.

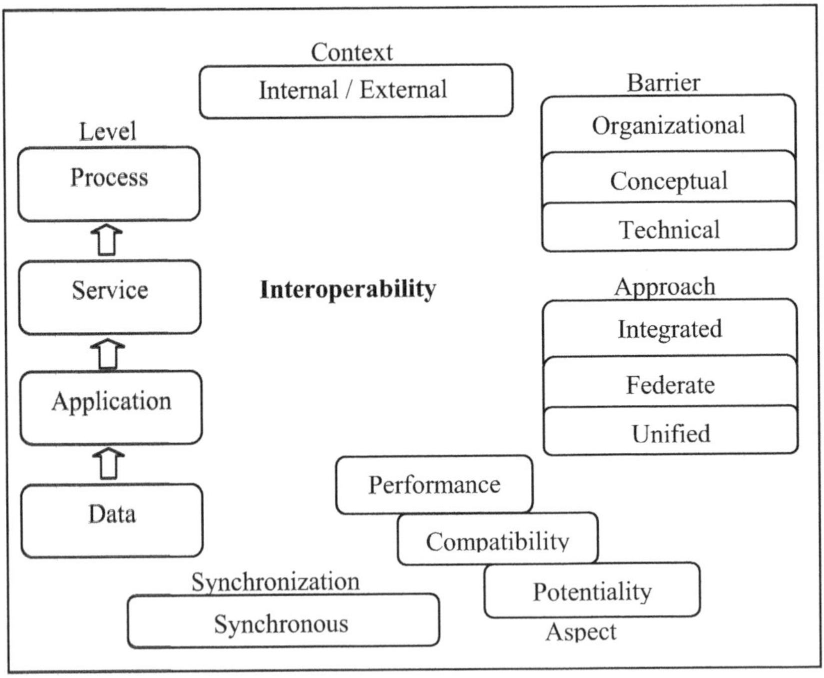

Figure 8.4. *Classification of interoperability*

8.3.1.1. *Approaches*

Three approaches are identified for achieving interoperability: integrated, unified or federated modes.

The integrated mode invites various partners to refer to a generic model for the interaction, and this in terms of data formats, protocols used, etc. The unified mode proposes a metamodel playing the role of middleware that everyone must work out in order to convert these interfaces according to this model for the accomplishment of the collaboration. As for the federated approach, it does not require full compliance with a specification or with a metamodel. But in this particular case, the various partners are invited to make additional efforts to remove ambiguities and accelerate interoperability.

8.3.1.2. *Fields of application*

Given the business process automation, organizations focus their efforts on the implementation of integrated management systems, each on a specific business. The next step is to make these various systems collaborate to improve internal operational performances. In this case, it involves the implementation of interoperability reflecting the internal character of the organization.

However, and from the perspective of the field of IS administration and management, the need to have two processes collaborate may go beyond the internal framework of an organization. In this case, interoperability assumes its external character (Guedria et al. 2008).

8.3.1.3. *Transactional mode*

There are two modes of interaction between systems. The first synchronous mode is established through transactions distributed on two or several heterogeneous transactional systems. Service-oriented solutions support this mode of communications. The second mode, referred to as asynchronous, describes the situations in which one part of the execution of operations is shifted, offline, to two or more heterogeneous transactional systems (Woolf 2009). The communications are at the basis of messages and events. The systems based on event-driven architecture illustrate this second operational mode.

8.3.1.4. *Operational aspect*

Similar to various qualities of the information system, there are three aspects related to interoperability: (i) the potential to interoperate as internal quality; (ii) the compatibility to interoperate referring to the external quality; and (iii) the operational performances corresponding to the quality during use.

8.3.1.5. *Barriers*

Most works concerning interoperability identify three classes of barriers that hinder the information exchange between the systems. These three categories of barriers identified and defined by Interop are classified into three levels (European Investment Fund 2008) as follows:

– The barriers known as *organizational*: they are related to organizational structures and management techniques used in various companies. More specifically, these barriers correspond to the definition of responsibility (who is responsible for what?) and of management of authorizations (who is authorized to do what?), as well as the compatibility of organizational structures (hierarchical compatibility) (Chen and Daclin 2006).

– The barriers known as *conceptual*: they correspond to differences in syntax and semantics between the exchanged entities. These differences concern, respectively, the coding of information, modeling and abstraction level of the latter, as, for example, a company model, and also the structuring of information for the exchange.

– The barriers known as *technological:* they concern the use of information materials to communicate and exchange information. These problems are related, for example, to incompatibilities at the level of platforms, infrastructures, operating systems, etc. From a purely technical point of view, these problems concern the protocols used to present, store, exchange, process and communicate data and information by means of information systems.

In order to overcome these barriers of interoperability, a significant number of works were conducted on the necessity of interoperability inside an organization and between organizations. According to a statistical study on this concept, the term "interoperability" has been used in scientific publications for nearly 50 years (Pirayesh Neghab 2014). This statistical study indicates that interoperability is a concept approached by several scientific domains. This may raise difficulties in understanding the concept, in the identification of barriers and finally in the search for appropriate indicators.

8.4. Evaluation of interoperability as a basis for the structural analysis of intelligent SoS

The evaluation of intelligent SoS remains a major challenge, and this is not only due to the complexity and the size of intelligent SoS. The interdependences between CS and the inherent interoperability influence, in most cases, the behavior of all SoS (Deleuze et al. 2013; Xia et al. 2016). This complexity is particularly critical in the interconnection and communication layer, which is responsible for establishing links and communication between the various systems composing intelligent SoS.

This work addresses the need for metrics to support the decision-making concerning the organization of structures inside intelligent SoS, by focusing specifically on the interconnection and communication layer. It is based on the structural analysis of intelligent SoS through the evaluation of interoperability in this layer. Its objective is to provide metrics for evaluating the effect of topology and of interdependences deteriorating the operation, both on the operability and on the system structure.

The structural analysis of the interconnection and communication is essential for understanding the effect of CS interdependences on the overall operation of intelligent SoS. By identifying the weak points and the potential congestion zones in this layer, intelligent SoS can take measures for reinforcing connectivity and ensuring a fluid communication between CS.

These metrics can also help in evaluating the impact of structural modifications of intelligent SoS on its overall performance, facilitating a better informed decision-making in terms of design and management of intelligent SoS.

Two metrics are proposed: the first one concerns the degree of barriers (DBs) of the interoperability deteriorating the performance of CS interdependences; the second one presents the Inefficiency of Exchanges (IE) within intelligent SoS.

8.4.1. *Quantification of interoperability barriers*

In order to quantify the barriers to interoperability, the matrix of the measure of interoperability compatibility, presented in the second chapter and cited article, is adopted. These barriers represent everything that blocks the path between the two systems.

Barriers are evaluated using a set of matrices, each of which is dedicated to a specific category of barriers. This evaluation is conducted for each aspect of barrier categories, based on four levels of interoperability. More precisely, every aspect of organizational, conceptual and technological barriers is evaluated based on the corresponding interoperability levels (Business, Processes, Service and Data).

Therefore, by assigning the weights p_i to each barrier by considering its relative importance and the criticality c_i of each barrier.

	Organizational barriers			
	Authorities/responsibilities		Organization	
Business	p_1	c_1	p_2	c_2
Processes	p_3	c_3	p_4	c_4
Service	p_5	c_5	p_6	c_6
Data	p_7	c_7	p_8	c_8

Table 8.2. *Organizational matrix*

It is important to note in Tables 8.2–8.4 the value of weight assigned to each barrier as a function of its relative importance with respect to the other barriers, as well as its criticality measure, reflecting the severity of the impact of the barrier on interoperability.

The weight p_i is a value ranging between 0 and 1, where 0 represents no importance and 1 represents a maximal importance.

The measure c_i is the criticality of a barrier. On a rating scale from 0 to 1, this measure c_i calculated for each barrier represents a numerical value indicating its level of criticality on the defined scale.

A barrier with 0 criticality measure is considered as having no impact on the interoperability, while a barrier with a criticality measure of 1 would have a more significant impact.

	Conceptual barriers			
	Syntactic		Semantic	
Business	p_1	c_1	p_2	c_2
Processes	p_3	c_3	p_4	c_4
Service	p_5	c_5	p_6	c_6
Data	p_7	c_7	p_8	c_8

Table 8.3. *Conceptual matrix*

The DB for a given barrier *i* denoted by "DB" is calculated according to formula [8.1].

$$DB_i = \frac{C_i \times P_i}{\sum_{i=1}^{n} P_i} \quad [8.1]$$

	Technological barriers			
	Platforms		Communication	
Business	p_1	c_1	p_2	c_2
Processes	p_3	c_3	p_4	c_4
Service	p_5	c_5	p_6	c_6
Data	p_7	c_7	p_8	c_8

Table 8.4. *Technological matrix*

Finally, since DB_i refers to the degrees of barriers, they are independent. We have chosen the weighted arithmetic mean to calculate the overall DBs described in formula [8.2].

$$DB = \frac{\sum_{i=1}^{n} DB_i}{\sum_{i=1}^{n} P_i} \quad [8.2]$$

8.4.2. *Inefficiency of communication*

A second measure we propose for completing the evaluation of interoperability is the inefficiency of communication (IC). It aims to evaluate the exchange in the overall system.

This is done through three aspects related to communication and its efficiency:

– the inefficiency rate of exchanges;

– the inability to interpret;

– noncompliance.

The inefficiency rate is the rate of unsuccessful exchanges. It is calculated by dividing the number of unsuccessful exchanges by the total number of exchanges, as indicated in formula [8.3], in this context, exchanges being all data, information, documents, etc. and exchanged through one interaction or one interdependence between at least two CS.

$$T_e = \frac{n_{uns}}{n_{tot}} \quad [8.3]$$

where:

T_e: represents the rate of inefficiency of exchanges;

n_{uns}: represents the number of unsuccessful exchanges;

n_{tot}: represents the total number of exchanges.

The rate of inability to interpret measures the degree of information, data or elements that are not interpreted when they are generated by a system of system (CS) and transferred to one or several other CS. This rate is calculated by dividing the number of information that was not interpreted to the total number of exchanges. The corresponding expression is presented in formula [8.4].

$$T_{int} = \frac{n_{int}}{n_{tot}} \quad [8.4]$$

where:

T_{int}: represents the inability to interpret;

n_{int}: represents the number of information, data or elements that were not interpreted, produced by a system and transferred to one or several other systems;

n_{tot}: represents the total number of exchanges.

Noncompliance represents the rate of noncompliant information, data, documents or anything that is generated by a system and transferred to one or several other systems. It is calculated by dividing the number of noncompliant information by the total number of received information (see formula [8.5])

$$T_{ncf} = \frac{n_{ncf}}{n_{tot}} \quad [8.5]$$

where:

T_{ncf}: represents the noncompliance;

n_{ncf}: represents the number of noncompliant exchanges;

n_{tot}: represents the total number of exchanges.

Finally, the IC represents the rate of global irregularity rates in the intelligent SoS. These are represented by the arithmetic mean of three aspects of the IC. The arithmetic mean is selected because the three aspects of inefficiency of the communication (rate of inefficiency of exchanges, inability to interpret and noncompliance) are independent, and there is no overlapping at any moment between them (see formula [8.6]).

$$IC = \frac{T_e + T_{int} + T_{ncf}}{3} \qquad [8.6]$$

In the context of an intelligent SoS, the previously mentioned metrics (DB and IC) are used in a formula to deduce a structural operability indicator (SOI). The SOI is used to evaluate the structural health of the global system using the rate of operability of interdependences.

The result of the SOI metrics is the rate of operability of a coalition of interdependent CS by considering its exposure to threats that are approached as barriers in addition to the inefficiency of the communication. The operability of SoS is calculated by a structural evaluation that is confirmed by a perspective of calculation of perturbations. The perturbations considered in our calculation are those aiming the interoperability. The calculation process must be applied in a similar manner to all interdependence based on the structure of the system, as indicated in formula [8.7]. It is important to note that the SOI reflects the rate of operability of interdependence.

$$SOI = 1 - \sqrt{DB \times EI} \qquad [8.7]$$

The objective of perturbations, both of the DB and of the IC, is to evaluate the ability of the interdependence to operate under the given circumstances, hence the name of the metrics. The higher the level of perturbations, the higher the risk of degradation of the performance and operability of the intelligent SoS. Therefore, SOI calculates the rate of operability among intelligent SoS taking into account the average of perturbations.

The formulation of the proposed structural operability equation takes into account the fact that the barriers and the IC may contribute to the lack of efficient operability of the intelligent SoS. This aims to develop a method for measuring interoperability in the context of intelligent SoS, converging to an SOI, which evaluates the structure of systems in terms of interoperability.

The following section presents an application of the explained theory, with a comparison between the proposal and the approach.

8.4.3. *Application to the case study*

This section presents an application of the theory to a case study. The studied sector is paper and cardboard manufacturing, represented by a coalition of companies gathering players in the paper–cardboard sector.

The information related to all aspects is collected through a questionnaire administered to the management of two players of the paper–cardboard sector.

The coalition of companies is considered as an SoS, in which the companies are represented by the CS and the interdependences represent the collaborations between companies.

It is important to note that information was gathered on the operation of the companies, and also on the perturbations and obstacles that hinder their interdependences. This has also helped in calculating the IC.

The information concerning the two players could not be disclosed, as they contain confidential commercial information. Tables will, however, be presented in order to identify the barriers to interoperability and reveal details on their nature through levels of interoperability.

Relying on gathered data, the matrices presented in Table 8.5 provide a detailed view of the weights and also of the barrier criticality measures, irrespective of their type: organizational, conceptual or technological. These barriers have been identified at each level of interoperability, namely activity, process, service and data, as soon as the two players started their interaction.

	Organizational barriers				**Conceptual barriers**				**Technological barriers**			
	Authorities/ responsibilities		Organization		Syntactic		Semantic		Platforms		Communication	
	pi	c_i	pi	c_i	Pi	c_i	pi	c_i	Pi	c_i	Pi	c_i
Business	0.8	0.2	0.2	0.2	0.2	0	0.2	0	0.3	0	0.2	0

Process	0.8	0.2	0.2	0.2	0.2	0	0.2	0	0.2	0	0.3	0
Service	0.7	0.4	0.4	0.4	0.4	0.1	0.4	0.2	0.9	0	0.9	0
Data	0.1	0.9	0.9	0.9	0.9	0.1	0.9	0.2	0.8	0.2	0.8	0.1

Table 8.5. *Matrices of interdependence between the two companies*

According to the defined method of barrier quantification, the global DBs are calculated. In this case, the DB is equal to 0.0951219512, which corresponds to 9.512% when expressed in percentage. This figure indicates the proportion of barriers with respect to the total number of evaluated interdependences.

Similarly, the IC is calculated by dividing the number of inefficient exchanges by the total number of exchanges between the components of the system of systems. For the considered interdependence, the IC is equal to 0.07223, or 7.223% when expressed in percentage. This number represents the proportion of inefficient exchanges in relation to all exchanges.

By combining these two measures, the SOI is obtained and is equal to 0.9171105644.

Finally, to express the rate of structural operability in percentage, we multiply SOI by 100, the result being 91.71105644%. This figure represents the level of structural operability of the evaluated system of systems, taking into account both the barriers and the inefficiency of the communication.

The information used was extracted from the same survey to complete the initial matrix suggested by Elmir and Bounabat (2010). In this approach, each studied aspect is evaluated to determine whether it has a barrier to interoperability. If a barrier is identified, it is assigned the value 1, otherwise, the value 0 is assigned to indicate the absence of barriers. The results obtained with this method are then summarized in Table 8.6.

The objective is to compare the results obtained with this approach with those previously obtained. An examination of both approaches shows that it is possible to better understand the differences in how the barriers are identified and quantified.

This comparison highlights the strong points and the limits of each method, thus offering a more complete perspective on the interoperability of the studied system of systems.

	Conceptual		Organizational		Technology	
	Syntactic	Semantic	Authorities	Organization	Platform	Communication
Business	0	0	0	0	0	0
Process	0	0	1	0	0	0
Service	1	1	1	1	0	0
Data	1	1	1	1	1	1

Table 8.6. *Application of the approach of Elmir and Bounabat (2010) to the interdependence between the two companies*

After applying this method, the global DBs are calculated to evaluate the cumulative impact of barriers on interoperability. In this case, the global DBs obtained are 0.4583333, which represents 45.833% when expressed in percentage. This figure indicates that nearly half of the evaluated aspects present barriers to interoperability.

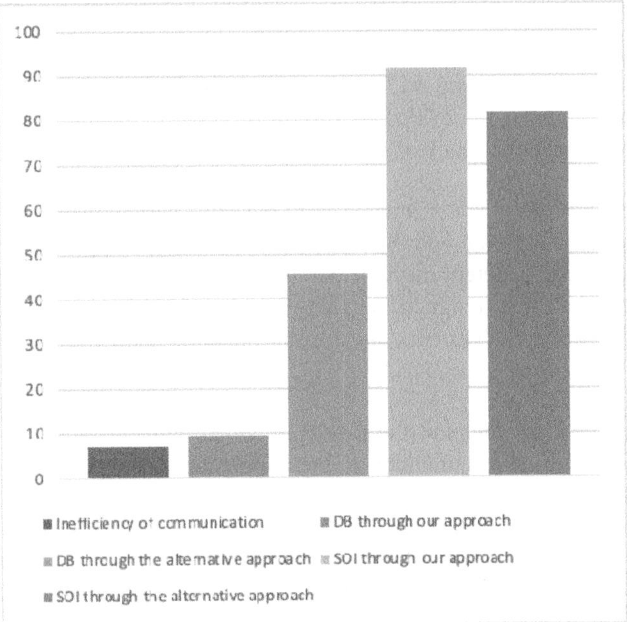

Figure 8.5. *Illustration of calculation results. For a color version of this figure, see www.iste.co.uk/elhami/applicationsofai.zip*

Although the rate of inefficiency of the communication is the same as in the previous evaluations, the value obtained for the SOI is equal to 0.8180510669.

Consequently, the rate of structural operability resulting from this approach is 81.80510669%. This figure represents the global capacity of the system of systems to operate efficiently and harmoniously, taking into account both the identified barriers and the IC between its components.

Figure 8.5 illustrates an intelligible view of the results. In this comparison of the two approaches, it is interesting to note that the proposed method assigns continuous values to the global DBs and the IC, while Elmir and Bounabat (2010) use a binary approach with values of 0 or 1 to indicate the presence or absence of barriers. This may lead to significant differences in the results obtained, as noted in the values of DB and IC.

In the first case, where the proposed method is used, the structural operability rate (SOI) is estimated at 91.71105644%. However, in the second case, in which the approach of Elmir and Bounabat (2010) is applied, the structural operability rate is slightly lower, at 81.80510669%. This difference may be attributed to the binary nature of the approach of Elmir and Bounabat (2010), which does not take into account the nuances in the severity of barriers and the IC.

Another important observation is that the proposed method seems to provide a more precise estimation of the structural operability, taking into account the continuous values for DB and IC. This facilitates a more nuanced evaluation of the impact of barriers and of the IC on the interoperability of SoS.

However, it is important to note that the approach of Elmir and Bounabat (2010) is easier to implement, as it uses a binary evaluation of barriers. This may be advantageous in situations in which a rapid qualitative evaluation is necessary.

In conclusion, the proposed method seems to offer a better precision in the estimation of structural operability of intelligent SoS taking into account the nuances in the barriers and the IC. However, the approach of Elmir and Bounabat (2010) may be more appropriate if a rapid qualitative evaluation is sufficient.

8.5. Conclusion

Finally, interoperability is a quality and not a feature of a system. It must be defined for at least two systems by the decision-makers during the design phase. Moreover, it can be considered as metrics dedicated to structural evaluation

(Han et al. 2012). This helps us to diagnose the interdependences forming the basic structure of SoS.

This chapter details an approach dedicated to the evaluation of structural operability of intelligent SoS. The proposed evaluation of structural operability can be used in many ways to enhance the performance and efficient operability of intelligent SoS. The user interface layer can provide operability indicators to users, who are thus able to monitor and understand the current state of the system and make informed decisions. By integrating these proposed metrics in the analysis and intelligence algorithms of layers for data collection, analysis and intelligence, these layers can identify the potential barriers to interoperability and to efficient communication between CS, thus anticipating the potential problems and taking preventive measures to solve them.

The metrics we proposed can also be used to support the coordination of actions of various CS to reach the global objectives of intelligent SoS at the level of the control and coordination layer. For example, if a critical barrier is identified, this layer may adjust the priorities to minimize its impact on the global performances of the system.

To summarize, the evaluation of structural operability can be integrated at each layer of the architecture of intelligent SoS to provide a continuous evaluation of the system's operability. This can be used for proactive decision-making and real-time adjustments to ensure an optimal operation of the intelligent SoS, despite the potential challenges related to interoperability between CS.

8.6. References

Abbott, N.J., Rönnbäck, L., Hansson, E. (2006). Astrocyte–endothelial interactions at the blood–brain barrier. *Nature Reviews Neuroscience*, 7(1), 41.

Abo Al-Kheer, A., Eid, M., Aoues, Y., El Hami, A., Kharmanda, M.G., Mouazen, A.M. (2011). Theoretical analysis of the spatial variability in tillage forces for fatigue analysis of tillage machines. *Journal of Terramechanics*, 48(4), 285–295.

Alrajeh, N.A., Elmir, B., Bounabat, B., El Hami, N. (2012). Interoperability optimization in healthcare collaboration networks. *Biomedizinische Technik – Biomedical Engineering*, 57(5), 403–411. doi: 10.1515/bmt-2011-0118.

Ammari, N., Ghallali, M., Abou El Kalam, A., El Hami, N., Ait Ouahman, A. (2014). Mobile security: Security mechanisms and protection of mobile applications. *Journal of Theoretical and Applied Information Technology*, 70(2), 302–315.

Ammari, N., El Hami, N., Ait El Mrabti, A., Abou El Kalam, A., Ait Ouahman, A. (2017). Securing mobile environment: Combination of MPSS framework with ALSD firewall. *Uncertainties and Reliability of Multiphysical Systems*, 1. doi: 10.21494/ISTE.OP.2017.0128.

Ben Abdessalem, A. and El Hami A. (2015). A probabilistic approach for optimising hydroformed structures using local surrogate models to control failures. *International Journal of Mechanical Sciences*, 96–97, 143–162.

Ben Abdessalem, A., Pagnacco, E., El Hami, A. (2013). Increasing the stability of T-shape tube hydroforming process under stochastic framework. *International Journal of Advanced Manufacturing Technology*, 69(5–8), 1343–1357.

Billaud, S., Daclin, N., Chapurlat, V. (2015). Interoperability as a key concept for the control and evolution of the system of systems (SoS). In *International IFIP Working Conference on Enterprise Interoperability*. Springer, Berlin, Heidelberg.

Boardman, J. and Sauser, B.J. (2006). The meaning of Systems of Systems. In *19th Annual System of Systems Engineering Conference*, Tacoma.

Bouchekourte, M. and El Hami, N. (2022). Optimization of equity allocations of institutional investors: Study of Moroccan case. *International Journal for Simulation and Multidisciplinary Design Optimization*, 13, 12.

Bouchekourte, M., Rhouas, S., El Hami, N. (2022). Minimizing the variance of the coverage ratio as an approach to optimize the exchange rate risk of Brent futures contracts. *International Journal for Simulation and Multidisciplinary Design Optimization*, 13, 17.

Bouyekhf, R., El Moudni, A., El Hami, A., Zerhouni, N., Ferney, M. (1996). Reduced order modelling of two-time-scale discrete non-linear systems. *Journal of the Franklin Institute*, 333(4), 499–512.

Bouyekhf, R., El Hami, A., El Moudni, A. (2001). Optimal control of a particular class of singularly perturbed nonlinear discrete-time systems. *IEEE Transactions on Automatic Control*, 46(7), 1097–1101.

Bradshaw, J.M., Feltovich, P.J., Jung, H., Kulkarni, S., Taysom, W., Uszok, A. (2003). Dimensions of adjustable autonomy and mixed-initiative interaction. In *International Workshop on Computational Autonomy*. Springer, Berlin, Heidelberg.

Bradshaw, J.M., Acquisti, A., Allen, J., Breedy, M.R., Bunch, L., Chambers, N., Feltovich, P., Galescu, L., Goodrich, M.A., Jeffers, R. et al. (2004). Teamwork-centered autonomy for extended human-agent interaction in space applications. In *AAAI 2004 Spring Symposium*, Palo Alto.

Chen, D. and Daclin, N. (2006). Framework for enterprise interoperability. In *Proc. of IFAC Workshop EI2N*, Bordeaux.

Chen, D., Doumeingts, G., Vernadat, F. (2008). Architectures for enterprise integration and interoperability: Past, present and future. *Computers in Industry*, 59(7), 647–659.

DeLaurentis, D. (2005). Understanding transportation as a system-of-systems design problem. In *43rd AIAA Aerospace Sciences Meeting and Exhibit*, Reno.

Deleuze, G., Léger, A., Piriou, P.-Y., Chabroux, S. (2013). Interoperability between a dynamic reliability modeling and a systems engineering process: Principles and case study. In *Embedded Real Time Software and Systems 2014 (ERTS22014)*, Toulouse.

DoD (2008). Systems engineering guide for systems of systems, volume 36. Report, Department of Defense, Washington, DC [Online]. Available at: https://acqnotes.com/wp-content/uploads/2014/09/DoD-Systems-Engineering-Guide-for-Systems-of-Systems-Aug-2008.pdf.

El Attaoui, A., Boukhafla, A., Rhouas, S., El Hami, N. (2023). Custom application programming interface data extractor applied to the Klarna e-commerce dataset. *Indonesian Journal of Electrical Engineering and Computer Science*, 31(3), 1624–1632.

El Hami, A. and Radi, B. (1996). Some decomposition methods in the analysis of repetitive structures. *Computers and Structures*, 58(5), 973–980.

El Hami, A., Lallement, G., Minotti, P., Cogan, S. (1993). Methods that combine finite group theory with component mode synthesis in the analysis of repetitive structures. *Computers and Structures*, 48(6), 975–982.

El Hami, N., Ellaia, R., Itmi, M. (2010). Hybrid evolutionary optimization algorithm MPSO-SA. *International Journal of Simulation and Multidisciplinary Optimization (IJSMDO)*, 4(1), 27–32.

El Hami, N., Ellaia, R., Itmi, M. (2011a). A new methodology for the reliability based particle swarm optimization with simulated annealing. *Advanced Materials Research*, 274, 107–118.

El Hami, N., Itmi, M., Ellaia, R. (2011b). Reliability-based design and heuristic optimization MPSO-SA of structures. *Advanced Materials Research*, 274, 95–106.

El Hami, N., Elmir, A., Elmir, B., Ellaia, R., Itmi, M., Bounabat, B. (2013). Simulation and optimization of interoperability planning. *Journal of Theoretical and Applied Information Technology*, 52(2).

El Hami, N., Itmi, M., El Hami, A. (2015a). Hybrid evolutionary optimization algorithm for structures. *Advanced Materials Research*, 1099, 102–109.

El Hami, N., Itmi, M., El Hami, A. (2015b). Simulation and optimization of an actuator. *Advanced Materials Research*, 1099, 110–119.

European Investment Fund (2008). Pan-European eGovernment Services. Draft Document as basis for EIF 2.0, European Investment Fund, Luxembourg.

Guedria, W., Naudet, Y., Chen, D. (2008). Interoperability maturity models – Survey and comparison. *Lecture Notes in Computer Science*, 5872/2009, 216–225.

Han, S.Y., Marais, K., DeLaurentis, D. (2012). Evaluating system of systems resilience using interdependency analysis. In *2012 IEEE International Conference on Systems, Man, and Cybernetics (SMC)*, Seoul.

IEEE (1990). Standard computer dictionary – A compilation of IEEE standard computer glossaries. Standard, IEEE Std 610. doi: 10.1109/IEEESTD.1991.106963.

Johnson, M., Bradshaw, J.M., Feltovich, P.J., Jonker, C.M., Van Riemsdijk, M.B., Sierhuis, M. (2014). Coactive design: Designing support for interdependence in joint activity. *Journal of Human-Robot Interaction*, 3(1), 43–69.

Kharmanda, G., Ibrahim, M.-H., Abo Al-kheer, A., Guerin, F., El Hami, A. (2014). Reliability-based design optimization of shank chisel plough using optimum safety factor strategy. *Computers and Electronics in Agriculture*, 109, 162–171.

Koulou, A., El Hami, N., Elmir, A., Hmina, N. (2017). A novel method for measuring interoperability of systems. *Uncertainties and Reliability of Multiphysical Systems*, 1. doi: 10.21494/ISTE.OP.2017.0125.

Koulou, A., Zemzami, M., El Hami, N., Elmir, A., Hmina, N. (2020). Optimization in collaborative information systems for an enhanced interoperability network. *International Journal for Simulation and Multidisciplinary Design Optimization*, 11, 2019021.

Lane, J.A. and Valerdi, R. (2011). System interoperability influence on system of systems engineering effort. In *Proceedings of the Conference on Systems Engineering Research* [Online]. Available at: https://www.researchgate.net/publication/268202801_System_Interoperability_Influence_on_System_of_Systems_Engineering_Effort.

Mabrouk, I.B. and El Hami, A. (2019). Effect of number of blades on the dynamic behavior of a Darrieus turbine geared transmission system. *Mechanical Systems and Signal Processing*, 121, 562–578.

Madni, A.M. and Sievers, M. (2014). System of systems integration: Key considerations and challenges. *Systems Engineering*, 17(3), 330–347.

Mahdi, A.E. (2020). Intelligent systems of systems: A comprehensive review. *Computers & Electrical Engineering*, 82, 106540.

Maier, M.W. (1998). Architecting principles for systems-of-systems. *Systems Engineering: The Journal of the International Council on Systems Engineering*, 1(4), 267–284.

Makhloufi, A., Aoues, Y., El Hami, A. (2016). Reliability based design optimization of wire bonding in power microelectronic devices. *Microsystem Technologies*, 22(12), 2737–2748.

Moro, T., El Hami, A., El Moudni, A. (2002). Reliability analysis of a mechanical contact between deformable solids. *Probabilistic Engineering Mechanics*, 17(3), 227–232.

Nielsen, C.B., Larsen, P.G., Fitzgerald, J., Woodcock, J., Peleska J. (2015). Systems of systems engineering: Basic concepts, model-based techniques, and research directions. *ACM Computing Surveys (CSUR)*, 48(2), 1–41.

Pingaud, H. (2009). Rationalité du développement de l'interopérabilité dans les organisations. In *8ème congrès international de Génie Industriel, CIGI09*. ENIT, Tarbes.

Pirayesh Neghab, A. (2014). Évaluation basée sur l'interopérabilité de la performance des collaborations dans le processus de conception. PhD Thesis, ENSAM, Paris.

Rhouas, S., Bouchekourte, M., El Hami, N. (2022). Optimization of the impact measurement of market structure on liquidity and volatility. *International Journal for Simulation and Multidisciplinary Design Optimization*, 13, 9.

Sage, A.P. and Cuppan, C.D. (2001). On the systems engineering and management of systems of systems and federations of systems. *Information Knowledge Systems Management*, 2(4), 325–345.

Tannahill, B.K. and Jamshidi, M. (2014). System of systems and big data analytics – Bridging the gap. *Computers & Electrical Engineering*, 40(1), 2–15.

Woolf, B. (2009). Event driven architecture and service oriented architecture. Paper, IBM Software Services, Winchester [Online]. Available at: https://research.ibm.com/haifa/Workshops/oopsla2006/present/w06_eda_woolf.pdf.

Xia, B., Zhao, Q., Dou, Y., Zhan, C. (2016). Robust system portfolio modeling and solving in complex system of systems construction. In *2016 35th Chinese Control Conference (CCC)*, 9573–9577. IEEE, Chengdu.

Zeine, A.T., El Hami, N., El Hami, A., Ouhimmou, S., Ellaia, R. (2017). Multiobjective optimization of trusses using backtracking search algorithm. *Incertitudes et fiabilité des systèmes multiphysiques*, 1. doi: 10.21494/ISTE.OP.2017.0121.

Zemzami, M., Makhloufi, A., El Hami, A., El Hami, N., Itmi, M., Hmina, N. (2017a). Applying a new parallelized version of PSO algorithm for electrical power transmission. In *IOP Conference Series: Materials Science and Engineering*, Kuala Lumpur. doi: 10.1088/1757-899X/205/1/012032.

Zemzami, M., El Hami, N., Itmi, M., Hmina, N. (2017b). A novel particle swarm optimization algorithm using dynamic neighborhood topology associated to parallel computation for complex optimization problems. *International Journal of Engineering & Technology (UAE)*, October.

Zemzami, M., El Hami, N., Itmi, M., Hmina, N. (2017c). An efficient hybrid model based on particle swarm optimization and simulated annealing algorithms for complex optimization problems. *International Journal of Engineering & Technology (UAE)*, September.

Zemzami, M., El Hami, N., Makhloufi, A., Itmi, M., Hmina, N. (2017d). Electrical power transmission optimization based on a new version of PSO algorithm. *Uncertainties and Reliability of Multiphysical Systems*, 1, 1–12.

Zemzami, M., Koulou, A., El Hami, N., Itmi, M., Hmina, N. (2017e). Optimizing interoperability using a new parallel model based on PSO algorithm. *International Journal of Engineering & Technology (UAE)*, October.

Zemzami, M., El Hami, N., Itmi, M., Hmina, N. (2019a). A modified particle swarm optimization algorithm linking dynamic neighborhood topology to parallel computation. *International Journal of Advanced Trends in Computer Science and Engineering,* 8(2), 112–118.

Zemzami, M., El Hami, N., Itmi, M., Hmina, N. (2019b). An evolutionary hybrid algorithm for complex optimization problems. *International Journal of Advanced Trends in Computer Science and Engineering*, 8(2), 126–133.

Zemzami, M., Koulou, A., El Hami, N., Itmi, M., Hmina, N. (2019c). Interoperability Optimization using a modified PSO algorithm. *International Journal of Advanced Trends in Computer Science and Engineering,* 8(2), 101–107.

List of Authors

Ali AKROUT
Laboratory of Mechanics, Modelling and Production (LA2MP)
National School of Engineers of Sfax (ENIS)
University of Sfax
Tunisia

Adnane CABANI
ESIGELEC
Saint-Étienne-du-Rouvray
France

Houcine CHAFOUK
ESIGELEC
Saint-Étienne-du-Rouvray
France

Anas EL ATTAOUI
ENSA-Kénitra
Université Ibn Tofail
Morocco

Abdelkhalak EL HAMI
Laboratory of Mechanics of Normandy (LMN)
INSA Rouen Normandie
France

Norelislam EL HAMI
ENSA-Kénitra
Université Ibn Tofail
Morocco

Abdelilah ELBAZZE
FST Settat
Morocco

Wajih GAFSI
Laboratory of Mechanics, Modelling and Production (LA2MP)
National School of Engineers of Sfax (ENIS)
University of Sfax
Tunisia

Sinda GHENAM
Laboratory of Mechanics of Normandy (LMN)
INSA Rouen Normandie
France
and
Laboratory of Mechanics, Modelling and Production (LA2MP)
National School of Engineers of Sfax (ENIS)
University of Sfax
Tunisia

Mohamed HADDAR
Laboratory of Mechanics, Modelling and Production (LA2MP)
National School of Engineers of Sfax (ENIS)
University of Sfax
Tunisia

Nabil HMINA
ENSA-Kénitra
Université Ibn Tofail
Morocco

Younes KOULOU
ENSA-Kénitra
Université Ibn Tofail
Morocco

Bouchaib RADI
FST Settat
Morocco

Hakima REDDAD
ENSA-Kénitra
Université Ibn Tofail
Morocco

Sara RHOUAS
ENSA-Kénitra
Université Ibn Tofail
Morocco

Maria ZEMZAMI
ENSAM
Université Mohamed V
Rabat
Morocco

Adnane ZOUBEIR
FST Settat
Morocco

Index

M, O

P, R

S, T

W, X

Other titles from

iSTE

in

Mechanical Engineering and Solid Mechanics

2024

LE VAN Anh
Contact in Structural Mechanics: A Weighted Residual Approach

LEROY Maurice
Rheology, Physical and Mechanical Behavior of Materials 2: Working at High Strain Rates, Forming, Sintering, Joining and Welding
Rheology, Physical and Mechanical Behavior of Materials 3: Rigidity and Resistance of Materials, Sizings, Pieces and Structures

SAOUAB Abdelghani, CHAMPMARTIN Stéphane, KHALID NACIRI Jaafar
Some Complex Phenomena in Fluid and Solid Mechanics
(Mathematical and Mechanical Engineering Set – Volume 13)

2023

EL HAMI Abdelkhalak
Uncertainty and Artificial Intelligence: Additive Manufacturing, Vibratory Control, Agro-composite, Mechatronics
(Artificial Intelligence in Mechanics Set – Volume 1)

EL HAMI Abdelkhalak, EID Mohamed
Reliability-based Modeling of System Performance
(Reliability of Multiphysical Systems Set – Volume 19)

EL HAMI Abdelkhalak, DELAUX David, GRZESKOWIAK Henri
Applied Reliability for Industry 1: Predictive Reliability for the Automobile, Aeronautics, Defense, Medical, Marine and Space Industries
(Reliability of Multiphysical Systems Set – Volume 16)
Applied Reliability for Industry 2: Experimental Reliability for the Automobile, Aeronautics, Defense, Medical, Marine and Space Industries
(Reliability of Multiphysical Systems Set – Volume 17)
Applied Reliability for Industry 3: Operational Reliability for the Automobile, Aeronautics, Defense, Medical, Marine and Space Industries
(Reliability of Multiphysical Systems Set – Volume 18)

LEROY Maurice
Rheology, Physical and Mechanical Behavior of Materials 1: Physical Mechanisms of Deformation and Dynamic Behavior

2022

BAYLE Franck
Product Maturity 1: Theoretical Principles and Industrial Applications
(Reliability of Multiphysical Systems Set – Volume 12)
Product Maturity 2: Principles and Illustrations
(Reliability of Multiphysical Systems Set – Volume 13)

EL HAMI Abdelkhalak, DELAUX David, GRZESKOWIAK Henri
Reliability and Physics-of-Healthy in Mechatronics
(Reliability of Multiphysical Systems Set – Volume 15)

LANNOY André
Reliability of Nuclear Power Plants: Methods, Data and Applications
(Reliability of Multiphysical Systems Set – Volume 14)

LEDOUX Michel, EL HAMI Abdelkhalak
Heat Transfer 3: Convection, Fundamentals and Monophasic Flows
(Mathematical and Mechanical Engineering Set – Volume 11)
Heat Transfer 4: Convection, Two-Phase Flows and Special Problems
(Mathematical and Mechanical Engineering Set – Volume 12)

PLANCHETTE Guy
Cindynics, The Science of Danger: A Wake-up Call
(Reliability of Multiphysical Systems Set – Volume 11)

2021

CHALLAMEL Noël, KAPLUNOV Julius, TAKEWAKI Izuru
Modern Trends in Structural and Solid Mechanics 1: Static and Stability
Modern Trends in Structural and Solid Mechanics 2: Vibrations
Modern Trends in Structural and Solid Mechanics 3: Non-deterministic Mechanics

DAHOO Pierre Richard, POUGNET Philippe, EL HAMI Abdelkhalak
Applications and Metrology at Nanometer Scale 1: Smart Materials, Electromagnetic Waves and Uncertainties
(Reliability of Multiphysical Systems Set – Volume 9)
Applications and Metrology at Nanometer Scale 2: Measurement Systems, Quantum Engineering and RBDO Method
(Reliability of Multiphysical Systems Set – Volume 10)

LEDOUX Michel, EL HAMI Abdelkhalak
Heat Transfer 1: Conduction
(Mathematical and Mechanical Engineering Set – Volume 9)
Heat Transfer 2: Radiative Transfer
(Mathematical and Mechanical Engineering Set – Volume 10)

2020

SALENÇON Jean
Elastoplastic Modeling

2019

BAYLE Franck
Reliability of Maintained Systems Subjected to Wear Failure Mechanisms: Theory and Applications
(Reliability of Multiphysical Systems Set – Volume 8)

BEN KAHLA Rabeb, BARKAOUI Abdelwahed, MERZOUKI Tarek
Finite Element Method and Medical Imaging Techniques in Bone Biomechanics
(Mathematical and Mechanical Engineering Set – Volume 8)

IONESCU Ioan R., QUEYREAU Sylvain, PICU Catalin R., SALMAN Oguz Umut
Mechanics and Physics of Solids at Micro- and Nano-Scales

LE VAN Anh, BOUZIDI Rabah
Lagrangian Mechanics: An Advanced Analytical Approach

MICHELITSCH Thomas, PÉREZ RIASCOS Alejandro, COLLET Bernard, NOWAKOWSKI Andrzej, NICOLLEAU Franck
Fractional Dynamics on Networks and Lattices

SALENÇON Jean
Viscoelastic Modeling for Structural Analysis

VÉNIZÉLOS Georges, EL HAMI Abdelkhalak
Movement Equations 5: Dynamics of a Set of Solids
(Non-deformable Solid Mechanics Set – Volume 5)

2018

BOREL Michel, VÉNIZÉLOS Georges
Movement Equations 4: Equilibriums and Small Movements
(Non-deformable Solid Mechanics Set – Volume 4)

FROSSARD Etienne
Granular Geomaterials Dissipative Mechanics: Theory and Applications in Civil Engineering

RADI Bouchaib, EL HAMI Abdelkhalak
Advanced Numerical Methods with Matlab® 1: Function Approximation and System Resolution
(Mathematical and Mechanical Engineering Set – Volume 6)
Advanced Numerical Methods with Matlab® 2: Resolution of Nonlinear, Differential and Partial Differential Equations
(Mathematical and Mechanical Engineering Set – Volume 7)

SALENÇON Jean
Virtual Work Approach to Mechanical Modeling

2017

BOREL Michel, VÉNIZÉLOS Georges
Movement Equations 2: Mathematical and Methodological Supplements
(Non-deformable Solid Mechanics Set – Volume 2)
Movement Equations 3: Dynamics and Fundamental Principle
(Non-deformable Solid Mechanics Set – Volume 3)

BOUVET Christophe
Mechanics of Aeronautical Solids, Materials and Structures
Mechanics of Aeronautical Composite Materials

BRANCHERIE Delphine, FEISSEL Pierre, BOUVIER Salima, IBRAHIMBEGOVIC Adnan
From Microstructure Investigations to Multiscale Modeling: Bridging the Gap

CHEBEL-MORELLO Brigitte, NICOD Jean-Marc, VARNIER Christophe
From Prognostics and Health Systems Management to Predictive Maintenance 2: Knowledge, Traceability and Decision
(Reliability of Multiphysical Systems Set – Volume 7)

EL HAMI Abdelkhalak, RADI Bouchaib
Dynamics of Large Structures and Inverse Problems
(Mathematical and Mechanical Engineering Set – Volume 5)
Fluid-Structure Interactions and Uncertainties: Ansys and Fluent Tools
(Reliability of Multiphysical Systems Set – Volume 6)

KHARMANDA Ghias, EL HAMI Abdelkhalak
Biomechanics: Optimization, Uncertainties and Reliability
(Reliability of Multiphysical Systems Set – Volume 5)

LEDOUX Michel, EL HAMI Abdelkhalak
Compressible Flow Propulsion and Digital Approaches in Fluid Mechanics
(Mathematical and Mechanical Engineering Set – Volume 4)
Fluid Mechanics: Analytical Methods
(Mathematical and Mechanical Engineering Set – Volume 3)

MORI Yvon
Mechanical Vibrations: Applications to Equipment

2016

BOREL Michel, VÉNIZÉLOS Georges
Movement Equations 1: Location, Kinematics and Kinetics
(Non-deformable Solid Mechanics Set – Volume 1)

BOYARD Nicolas
Heat Transfer in Polymer Composite Materials

CARDON Alain, ITMI Mhamed
New Autonomous Systems
(Reliability of Multiphysical Systems Set – Volume 1)

DAHOO Pierre Richard, POUGNET Philippe, EL HAMI Abdelkhalak
Nanometer-scale Defect Detection Using Polarized Light
(Reliability of Multiphysical Systems Set – Volume 2)

DE SAXCÉ Géry, VALLÉE Claude
Galilean Mechanics and Thermodynamics of Continua

DORMIEUX Luc, KONDO Djimédo
Micromechanics of Fracture and Damage
(Micromechanics Set – Volume 1)

EL HAMI Abdelkhalak, RADI Bouchaib
Stochastic Dynamics of Structures
(Mathematical and Mechanical Engineering Set – Volume 2)

GOURIVEAU Rafael, MEDJAHER Kamal, ZERHOUNI Noureddine
From Prognostics and Health Systems Management to Predictive Maintenance 1: Monitoring and Prognostics
(Reliability of Multiphysical Systems Set – Volume 4)

KHARMANDA Ghias, EL HAMI Abdelkhalak
Reliability in Biomechanics
(Reliability of Multiphysical Systems Set –Volume 3)

MOLIMARD Jérôme
Experimental Mechanics of Solids and Structures

RADI Bouchaib, EL HAMI Abdelkhalak
Material Forming Processes: Simulation, Drawing, Hydroforming and Additive Manufacturing
(Mathematical and Mechanical Engineering Set – Volume 1)

2015

KARLIČIĆ Danilo, MURMU Tony, ADHIKARI Sondipon, MCCARTHY Michael
Non-local Structural Mechanics

SAB Karam, LEBÉE Arthur
Homogenization of Heterogeneous Thin and Thick Plates

2014

ATANACKOVIC M. Teodor, PILIPOVIC Stevan, STANKOVIC Bogoljub, ZORICA Dusan
Fractional Calculus with Applications in Mechanics: Vibrations and Diffusion Processes
Fractional Calculus with Applications in Mechanics: Wave Propagation, Impact and Variational Principles

CIBLAC Thierry, MOREL Jean-Claude
Sustainable Masonry: Stability and Behavior of Structures

ILANKO Sinniah, MONTERRUBIO Luis E., MOCHIDA Yusuke
The Rayleigh–Ritz Method for Structural Analysis

LALANNE Christian
Mechanical Vibration and Shock Analysis – 5-volume series – 3rd edition
Sinusoidal Vibration – Volume 1
Mechanical Shock – Volume 2
Random Vibration – Volume 3
Fatigue Damage – Volume 4
Specification Development – Volume 5

LEMAIRE Maurice
Uncertainty and Mechanics

2013

ADHIKARI Sondipon
Structural Dynamic Analysis with Generalized Damping Models: Analysis

ADHIKARI Sondipon
Structural Dynamic Analysis with Generalized Damping Models: Identification

BAILLY Patrice
Materials and Structures under Shock and Impact

BASTIEN Jérôme, BERNARDIN Frédéric, LAMARQUE Claude-Henri
Non-smooth Deterministic or Stochastic Discrete Dynamical Systems: Applications to Models with Friction or Impact

EL HAMI Abdelkhalak, RADI Bouchaib
Uncertainty and Optimization in Structural Mechanics

KIRILLOV Oleg N., PELINOVSKY Dmitry E.
Nonlinear Physical Systems: Spectral Analysis, Stability and Bifurcations

LUONGO Angelo, ZULLI Daniele
Mathematical Models of Beams and Cables

SALENÇON Jean
Yield Design

2012

DAVIM J. Paulo
Mechanical Engineering Education

DUPEUX Michel, BRACCINI Muriel
Mechanics of Solid Interfaces

ELISHAKOFF Isaac *et al.*
Carbon Nanotubes and Nanosensors: Vibration, Buckling and Ballistic Impact

GRÉDIAC Michel, HILD François
Full-Field Measurements and Identification in Solid Mechanics

GROUS Ammar
Fracture Mechanics – 3-volume series
Analysis of Reliability and Quality Control – Volume 1
Applied Reliability – Volume 2
Applied Quality Control – Volume 3

RECHO Naman
Fracture Mechanics and Crack Growth

2011

KRYSINSKI Tomasz, MALBURET François
Mechanical Instability

SOUSTELLE Michel
An Introduction to Chemical Kinetics

2010

BREITKOPF Piotr, FILOMENO COELHO Rajan
Multidisciplinary Design Optimization in Computational Mechanics

DAVIM J. Paulo
Biotribology

PAULTRE Patrick
Dynamics of Structures

SOUSTELLE Michel
Handbook of Heterogenous Kinetics

2009

BERLIOZ Alain, TROMPETTE Philippe
Solid Mechanics using the Finite Element Method

LEMAIRE Maurice
Structural Reliability

2007

GIRARD Alain, ROY Nicolas
Structural Dynamics in Industry

GUINEBRETIÈRE René
X-ray Diffraction by Polycrystalline Materials

KRYSINSKI Tomasz, MALBURET François
Mechanical Vibrations

KUNDU Tribikram
Advanced Ultrasonic Methods for Material and Structure Inspection

SIH George C., NAÏT-ABDELAZIZ Moussa, VU-KHANH Toan
Particle and Continuum Aspects of Mesomechanics